SOME DETAILS OF WATER-WORKS CONSTRUCTION

NOYES PRESS SERIES IN HISTORY OF TECHNOLOGY

The books published in the History of Technology Series are reprints of important works published in the eighteenth and nineteenth centuries. Most of them are of American origin, however some were published in Great Britain, or are early translations of European works.

In addition to describing historical technological devices and processes, many of the books give an insight into the relationship of early technology to the culture of the day.

1. CONSTRUCTION OF MILL DAMS
   by James Leffel, 1881

2. SOME DETAILS OF WATER-WORKS CONSTRUCTION
   by William R. Billings, 1898

3. THE MANUFACTURE OF LIQUORS AND PRESERVES
   by J. de Brevans, 1893

4. THE MANUFACTURE OF PORCELAIN AND GLASS
   by Dionysius Lardner, 1832

SOME DETAILS OF

WATER-WORKS CONSTRUCTION

History of Technology Vol. No. 2

William R. Billings

NOYES PRESS
Noyes Building
Park Ridge, New Jersey 07656, U.S.A.

A Noyes Press Reprint Edition

This edition of SOME DETAILS OF WATER-WORKS CONSTRUCTION is an unabridged republication of the first edition published in 1898 by McGraw Publishing Co.

Library of Congress Catalog Card Number: 72-80387
ISBN: 0-8155-5006-5
Printed in the United States

FOREWORD
1972

This book (third edition), published in 1898 by the McGraw Publishing Company, was based on a series of articles included in the ENGINEERING RECORD, actually written ten years earlier in 1888. The author, William R. Billings, was Superintendent of the Water-Works at Taunton, Mass. from 1879 to 1888.

This book is a series of practical articles, based on actual experience of the author, and offers considerable insight into water-works practice in the mid-nineteenth century.

Chapters include:

1. The Distribution System
2. Field Work
3. Trenching and Pipe-Laying
4. Joint-Making
5. Hydrants, Gates, and Specials
6. Service-Pipes
7. Meters
8. Construction Details
9. Costs

SOME DETAILS

OF

Water-Works Construction.

BY

WILLIAM R. BILLINGS,

Superintendent Water-Works at Taunton, Mass.
(From 1879 to 1888.)

WITH

ILLUSTRATIONS FROM SKETCHES BY THE AUTHOR.

THIRD EDITION.

NEW YORK
McGRAW PUBLISHING COMPANY
114 LIBERTY STREET
1898

INTRODUCTORY NOTE.

SOME questions addressed to the Editor of THE ENGINEERING RECORD by persons in the employ of new water–works indicated that a short series of practical articles on the Details of Constructing a Water–Works Plant would be of value; and, at the suggestion of the Editor, the preparation of these papers was undertaken for the columns of that journal. The task has been an easy and agreeable one, and now, in a more convenient form than is afforded by the columns of the paper, these notes of actual experience are offered to the water-works fraternity, with the belief that they may be of assistance to beginners and of some interest to all.

TABLE OF CONTENTS.

TABLE OF CONTENTS.

LIST OF ILLUSTRATIONS.

CHAPTER I.

THE DISTRIBUTING SYSTEM.

Materials — Salt-Glazed Clay — Cast Iron — Cement-Lined Wrought-Iron—Thickness of Sheet Metal—Methods of Lining — List of Tools — Tool-Box — Derrick — Calking Tools—Furnace—Transportation—Handling Pipe—Cost of Carting—Distributing Pipe.

IN considering the subject of which this book treats, it will be the writer's endeavor to be brief and practical.

He assumes that those for whom these papers will have the most interest have had little or no experience in actual construction, and desire information and suggestion upon the simplest details.

MAIN PIPES.

Materials.—Cast iron, wrought iron with cement or with a protecting coating by some special process, wood, and steel are the materials used in making pipes for the distributing systems of town and city water-supplies.

Salt-glazed vitrified clay pipes have been used by Mr. Stephen E. Babcock, C. E., of Little Falls, N. Y., in that village, and also at Amsterdam and Johnstown in the same

State, for conduits in gravity systems. At Little Falls the conduit is over 30,000 feet in length and is mainly of 18 and 20 inch pipe. The low first cost of clay pipe would certainly entitle its claims to careful investigation in planning a low-pressure gravity system of supply. Mr. Babcock has prepared a very elaborate set of specifications for furnishing and laying this pipe, which would be of value to any one who wished to use it.

The writer frankly acknowledges a preference for cast-iron pipe for all but special cases. He is not unmindful of the fact that the town of Plymouth, Mass., after an experience of thirty years, is this summer (1887) extending its distributing system by adding 20,000 feet of 4 to 16 inch cement-lined wrought-iron pipe; nor that the town of Dedham, Mass., has had no reason thus far to regret that its water-mains are of this material. Without going further, the cities of Fitchburg and Worcester, Mass., seem to offer experience with this sort of pipe to justify the opinion that the chances for poor work and poor material are greater with it than with cast iron, and the advocates of cement-lined pipe admit, I think, that honest and skillful work is indispensable to the success of this method. We must admit that when made and laid upon honor, cement-lined pipe has an advantage over cast iron in not reducing its original diameter by incrustations, nor is the "advantage out' when we reply that the cleaning machine of Mr. Keating or of Mr. Sweeney may be used to restore tuberculated iron pipe to its original diameter, for the application of these machines cannot be effected without expense.

With its acknowledged advantages of strength and ease in laying, cast-iron pipe is heavy and, in its larger sizes, expensive to handle. This limits the length in which sections of it can be used, and so does not permit of any reduction in the number of joints to the mile.

In the effort to produce something which should be free from these disadvantages of cast-iron pipe, wrought-iron pipe treated by a protective process is now upon the market, and has been introduced to a limited extent. Of this it is fair to say that it is still on trial, and some time must yet elapse before its durability can be said to be proven.

Of wood, the writer has no knowledge by actual experience, but its use seems to be limited to a small territory in the West.

Unlike cast-iron pipe, which is bought ready for use, cement-lined pipe is put together in part at some convenient yard or shop in the town which is to use it, and its final construction is carried on in the trench where it is to lie.

The foundation of this sort of pipe is a sheet-iron drum nine feet in length, made in three sections in the 16-inch and larger sizes, and in single sheets in the 4 to 14 inch diameters.

The thickness of metal varies with the sizes ; for example we may use,

For 4-inch	pipe,	metal of	21	Birmingham	Gauge,
" 6	"	"	19	"	"
" 10	"	"	17	"	"
" 12	"	"	15	"	"
" 14	"	"	14	"	"
" 16	"	"	14	"	"

with double-riveted seams, using 12-pound rivets for 16, 14, and 12 inch pipe, 10-pound for 10-inch, 6-pound for 6-inch, and 5-pound for 4-inch pipes.

The first step in the making of this sort of pipe is the putting into these drums a lining ½ to ¾ of an inch in thick-

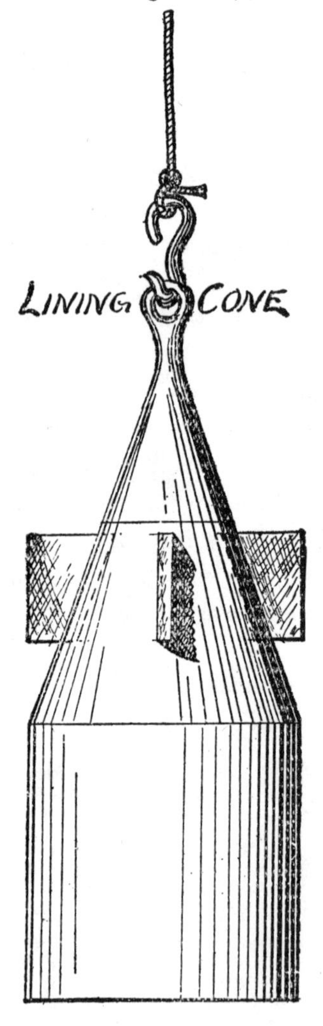

FIGURE 1.

ness of cement mortar mixed sand and cement half and half.

The pipe is placed on end over a hole in a low platform, and a lining cone is let down into it from a crane, a derrick, or a simple windlass, and drops through the hole in the platform just far enough to allow the pipe to be entirely filled at its lower end with the mortar. Enough mortar is then shoveled into the top of the pipe from a high platform to make the lining, and the cone is drawn slowly through. The surplus cement as it falls over the top during the upward movement of the cone is shoveled back into the mixing-box, or into another pipe if there be one at hand ready for lining, but no cement that has once set is fit to be used again. After the cone is drawn the pipe should stand 20 minutes or more before it is moved; it is then taken to the grouting table, the ends scraped, and the whole surface examined for defects. If at any points the cement has settled into wrinkles these should be scraped down, and any holes filled with pure cement.

With platforms and swinging crane arranged to place ten pipes on end at once for lining, eight men can fill 100 14-inch pipes in a day, and three men more can grout and patch them.

The grout can be poured in with a dipper, and then spread by rolling the pipe and applying from each end common dust-brushes fastened to long handles.

Before applying the grout the lining is brushed with water, using the long-handled brushes.

The lining cones are made either of cast iron or of sheet-metal, but if the latter is used they must be filled with cement to give them weight.

LIST OF TOOLS.

Whatever be the material chosen for main pipe, the trenching tools will be the same. In the matter of pipe cutting and jointing, cast iron and wrought iron call for very different treatment and appliances. During days which are too stormy for work and over night, all tools should be securely packed in tool-boxes, which may be built according to the following sketch :

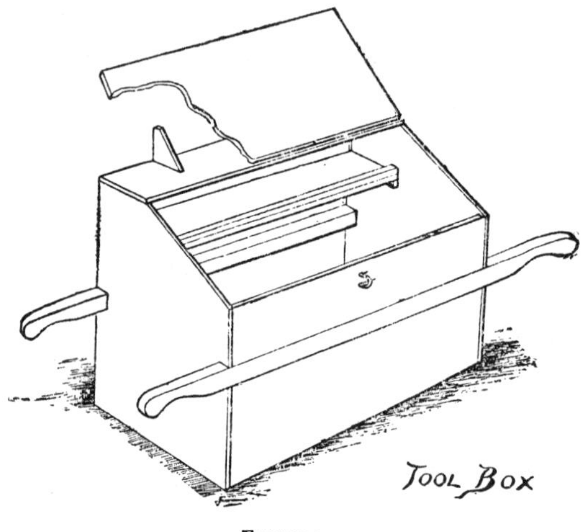

FIGURE 2.

The same carpenter who makes the boxes can make also a derrick after the sketch given in Figure 3, which will be found strong enough for pipes weighing a ton and easy to handle as soon as three men get the knack of carrying it.

This should be made of straight-grained 4x4 sticks, 14½ feet long, held together at the top by a 1-inch bolt. The link of ⅞ round iron drops one foot, and ⅜ carriage-bolts should be put through the end of the sticks to keep them from splitting. The large cleat on the right is to be bolted on with two ⅜ carriage-bolts about 20 inches from the bottom of the leg, and a hard-wood pin driven in about the same distance from the bottom of each of the other legs. For pipes larger than 20 inches a 4-leg derrick with a windlass may be found more convenient.

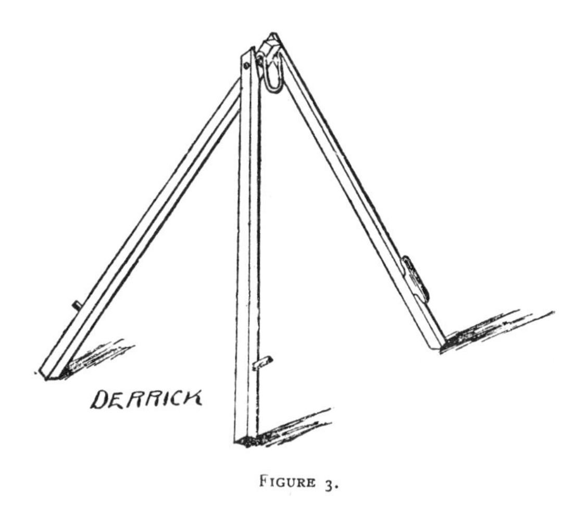

FIGURE 3.

For 6-inch pipe two 8-inch double blocks will give power enough, but for 16-inch a quadruple and triple block in combination will be needed.

The number of picks and shovels required depends, of course, upon the number of men that are to be employed. One shovel to a man is enough, but if the digging is likely to be hard, double the number of picks will not be too many, to allow time for sharpening. A shovel with a welded strap does better work than one in which the strap is riveted, and for anything but scraping up from a platform, a round point is better than a square point.

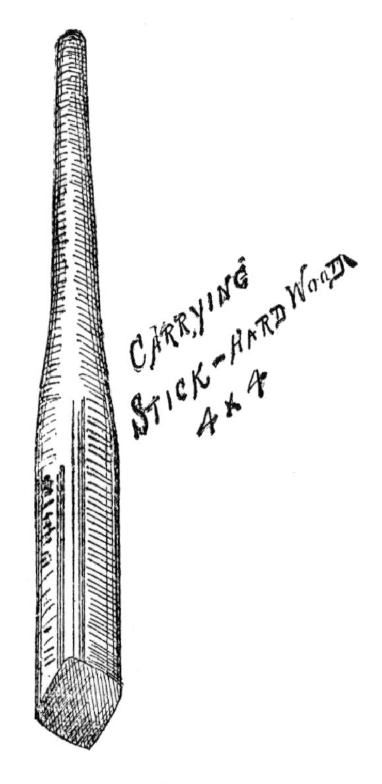

FIGURE 4.

Provide three, four, or half a dozen steel crowbars 5½ to 6 feet long, 2 or 3 sledges weighing, say, 10, 15, and 20 pounds, and 2 tunneling bars, if the digging will permit of this sort of work. The tunneling bars are easily made by welding on to a piece of 1-inch pipe 8 or 10 feet long a chisel-shaped piece of steel 2 or 3 inches wide.

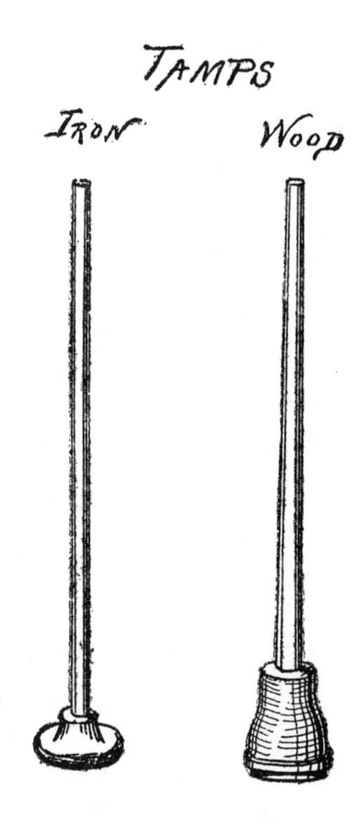

FIGURE 5.

For ledge-work, drills made of 1⅛-inch octagon steel, forged to cut a 1¼-inch hole, with sledges weighing 6 to 8 pounds, and a spoon for scooping out the dust and drillings, will be required.

Carrying sticks for lifting 4, 6, and 8 inch pipe, of the shape shown in the sketch Figure 4 (page 16), are useful; larger sizes of pipe are handled more easily by rolling.

FIGURE 6.

Skids 6 feet long of 2x4, 4x4, or of 4x6 spruce, according to the weight of the pipe, will be needed to throw across the trench.

When water is not available for back-filling some kind of tamp will be needed, and sketches of two patterns are given on page 17 (Figure 5).

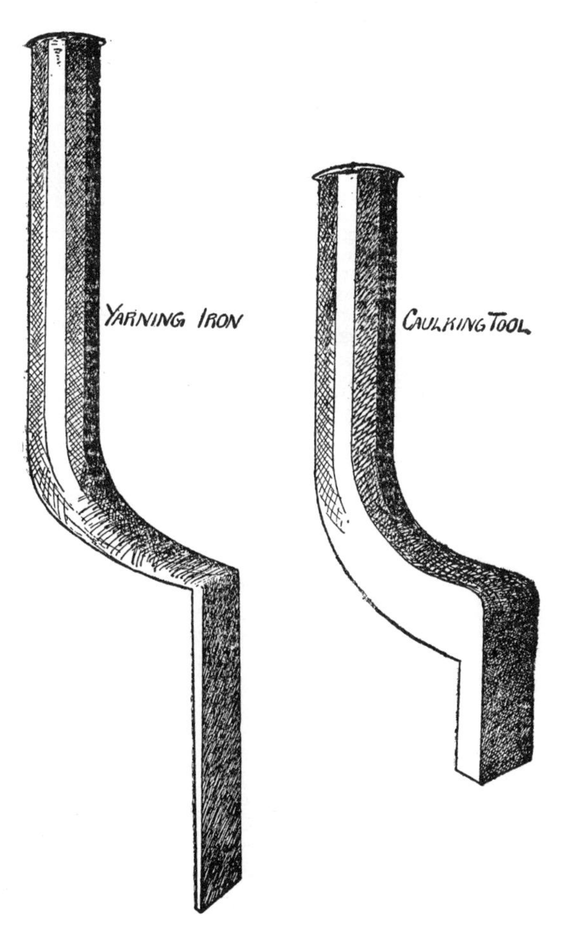

FIGURE 7.

If any considerable amount of rock-work is expected, either as ledge or boulders, a second derrick will be needed, with some spare rope and a few pieces of chain.

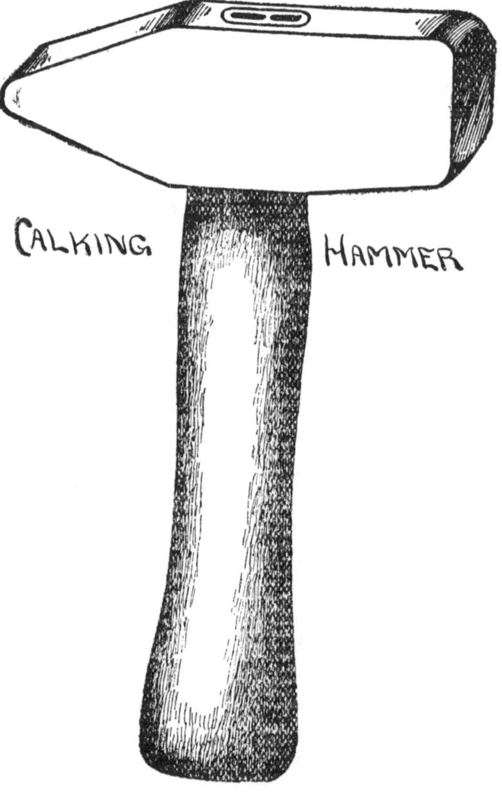

FIGURE 8.

For cutting cast-iron pipe provide two or three long-handled chisels, such as blacksmiths use for cutting off cold iron (see Figure 6, page 18), and a pair of light sledges or striking hammers. For cutting wrought-iron pipe boiler-makers' chisels and hammers are the proper tools.

For making lead joints in cast-iron pipe, yarning and calking tools and short-handled calking hammer. One yarning-

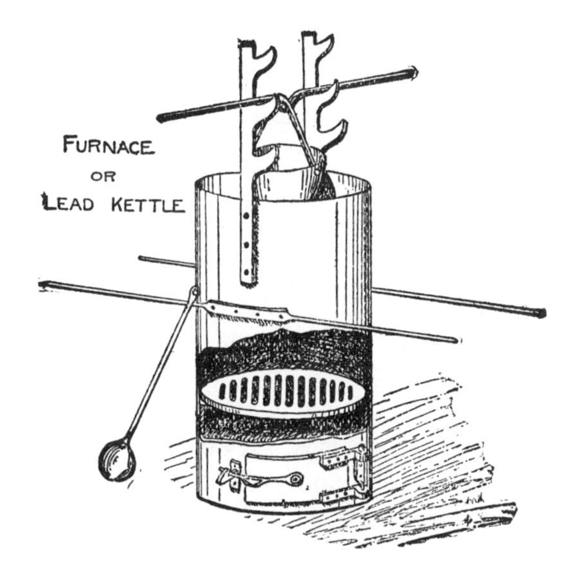

FIGURE 9.

iron and four calking tools varying in thickness from 1/8 to 3/8 of an inch make a convenient set.

A furnace or lead-kettle of the pattern indicated in Figure 9 is common among water-works contractors. There should be a second door opening on to the grate at the point on the sketch where the shell is broken away to show the interior.

TRANSPORTATION.

Before the arrival of the pipe arrangements should be made to have men and teams ready to begin work at a few hours' notice; for, as a rule, vessel captains and railroad companies are in a hurry to be rid of their cargoes. Some trustworthy man should be selected to oversee the unloading and keep tally. In cast-iron pipe it is customary to mark the weight of each piece with white paint inside the bell, and if a memorandum is made of the weight of each piece as it leaves the car or vessels, the pipes will be counted and a check on the weight given in the bill of lading will be obtained.

The pipes may be piled up on the wharf, or taken directly from the cars on to the drays or low gears that are to cart them. If they are to be put directly upon the drays little or no blocking will be needed. Strong and careful men with carrying sticks, and some skids in the absence of a derrick, will soon discover the easiest method of handling the pipe and avoiding shocks and blows. If the pipe is coming out of a vessel and is to be piled up on the wharf, 2x4 spruce sticks in market lengths should be placed between the tiers, and strong skids used to roll the pipe from the deck ashore, and blocking be freely applied to prevent bunting, striking, or rolling. In the experience of the writer, six active and fearless men easily took 16, 6, and 4 inch pipe from a vessel and piled it securely on a wharf faster than the crew could get it out of the hold with a steam derrick.

In carting cast-iron pipe convenience and necessity will determine the kind of vehicle to be used, but in carting cement-lined pipe it is well to insist that the wagon shall have springs that the chances for cracking the cement lining may be reduced. The cost of carting must vary so much with circumstances that the writer can do no more than quote some figures from his own experience. Three bids were received in the spring of 1887 for carting an average distance of about two miles over good roads and streets with no steep grades:

680 tons of 135-lb. 16-inch pipe.
100 " 34-lb. 6-inch pipe.
100 " 20-lb. 4-inch pipe.
One of $1 per ton gross.
One of 67$\frac{2}{5}$ cents per ton gross.
One of 64 cents per ton gross.

At the lowest figure the teamster appears to be satisfied that he is making a fair profit, but his horses and men are working hard for it.

When the town of Middleboro, Mass., constructed its water-works the writer was informed that the carting was done for fifty cents per ton, but the average hauls were short and the roads good.

Considerable judgment is required on the part of the teamsters who deliver the pipe on the street to distribute it so that it will not fall short or run over in laying, so that it will not cause excessive risk to night travelers while it is awaiting the coming of the workmen, so that it will not be in the way of entrances to private estates or of merchants whose teams wish to receive or discharge goods. If circumstance will permit, it will save time for the pipe-layers to have the pipe laid on the street with all the bells pointing one way, and that in the direction of the movement of the gang. This with reference to bell and spigot pipe ; with other patterns this condition does not exist.

In directing the teamsters on which side of the street to deliver the pipe, consider on which side of the trench the bulk of the dirt is to be thrown, and have the pipe dropped on the side opposite to that, and thus avoid having to lift the pipe over an embankment of loose earth.

CHAPTER II.

FIELD WORK.

Engineering or None—Pipe Plans—Special Pipe—Laying Out a Line—Width and Depth of Trench—Time-Keeping Book—Disposition of Dirt—Tunneling—Street-Piling.

IT is well understood by the readers of THE ENGINEERING AND BUILDING RECORD that the best preparation for any considerable amount of main-pipe laying is found in a careful survey of the proposed line, which shall take note of every feature which is likely to affect the work. Cross streets or roads, existing or proposed, brooks, bridges, drains, culverts, sewers, gas-pipes, and old water-mains, if there be any, should be indicated on plan and profile, and forethought given to schemes for avoiding and overcoming evident obstacles.

Let me warn the novice that, in spite of his most earnest forethought, obstacles that could hardly be foreseen even by one of experience will almost certainly arise, and he can at best only strive to reduce the number of the unexpected difficulties.

The need for laying pipes to line and grade is an imperative one on the main line from a reservoir in a gravity system ; is almost as necessary with any main larger than ten or twelve inches, though perhaps less important in the smaller pipes through the streets of a town.

A town may build a respectable system of water-works with a wonderfully small amount of engineering, but money saved at the outset in this way is generally expended at a later date in correcting blunders and repairing defective work. The writer calls to mind at this moment an instance in which a defective length of pipe which was made a part of a submerged river-crossing has since caused an expenditure of not less than $2,000 at different times for repairs; enough to have paid for a reasonable amount of engineering and thorough inspection.

Let us suppose that full surveys and drawings have been made; in what form, then, shall they be put, so as to be intelligible to the foreman in charge of the gang? If an assistant engineer is constantly on the trench, he may not need a full drawing; his own notes made at the office may be sufficient, but this arrangement is not always practicable. We give herewith Figure 10, a sketch copied from blue prints used by Mr. R. C. P. Coggeshall in his work at New Bedford, Mass. These sheets are not large, 10x15 inches or less, and are given to the foreman a few days before the beginning of the work, so that he may get the gates, hydrants, and specials on the ground in advance of the digging. The writer has followed essentially the same plan, but in his drawings no attempt is made to show the form of the special castings. Single or double lines, with the names of the castings and size of gates, the whole drawn to scale of forty feet to one inch, are used as shown in Figure 11, page 29.

If a draughtsman is available the first method is certainly to be preferred, but if one must be his own engineer, super-

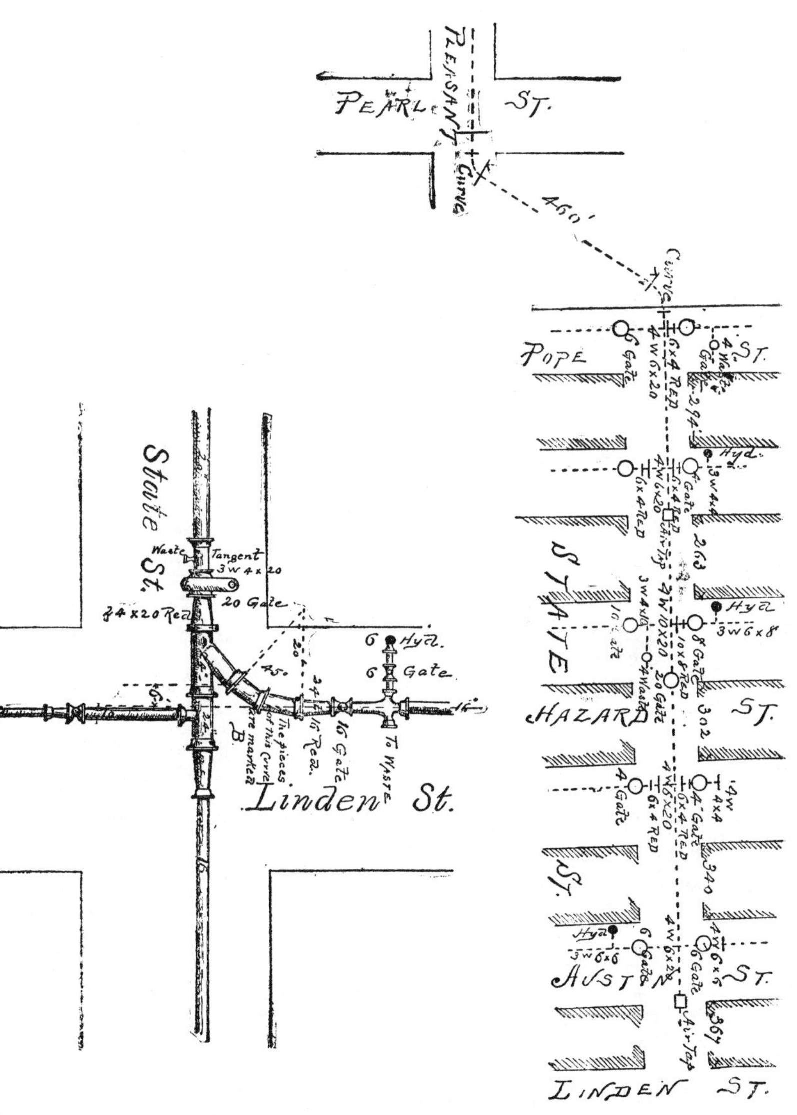
PEARL ST.
PLEASANT
CURVE
460'
POPE ST.
HAZARD ST.
AUSTIN ST.
LINDEN ST.
STATE ST.
State St.
Linden St.
Waste
Tangent
3 W 4 x 20
20 Gate
24 x 20 Red.
6 Hyd.
6 Gate
45°
16 Gate
16 Red.
To Waste
The pieces of this Curve are marked B.

FIGURE 10.

intendent, and draughtsman, as is often the case in small towns, the second method has its advantages.

These pipe plans represent the best practice, but if the earth could be thrown off some main-pipe systems, as the valet of Frederick the Great used to throw the bed-clothes off his master in the morning, the easy curves and special angles which the foregoing plans provide would not be found.

I once heard a man of wide experience in handling pipe say that he could lay cast-iron pipe in the crookedest town that was ever laid out on the cow-paths, if he had single branches and plenty of pipe. Such work is not to be commended, but it has been, and probably will be, done.

It would be outside the scope of these papers to describe methods which may be employed in locating a pipe-line and staking out the curves and angles according to railroad practice, but some beginner may be glad to receive suggestions as to simple working methods. When he is given a gang of men, a quantity of straight pipe, and told to lay a main on B Street from a given point to a branch on S Avenue, his instructions will probably include the location of the branch with more or less accuracy, but on reaching the scene of operations he may find that this branch points all askew for B Street, and the street itself straggles along to a junction with the avenue in a tangential, uncertain sort of way that is more picturesque than satisfactory.

The first comforting fact is that we *can* swing around the sharpest curve which is likely to present itself by cutting the pipe into short pieces and making the joints as one-sided as we dare.

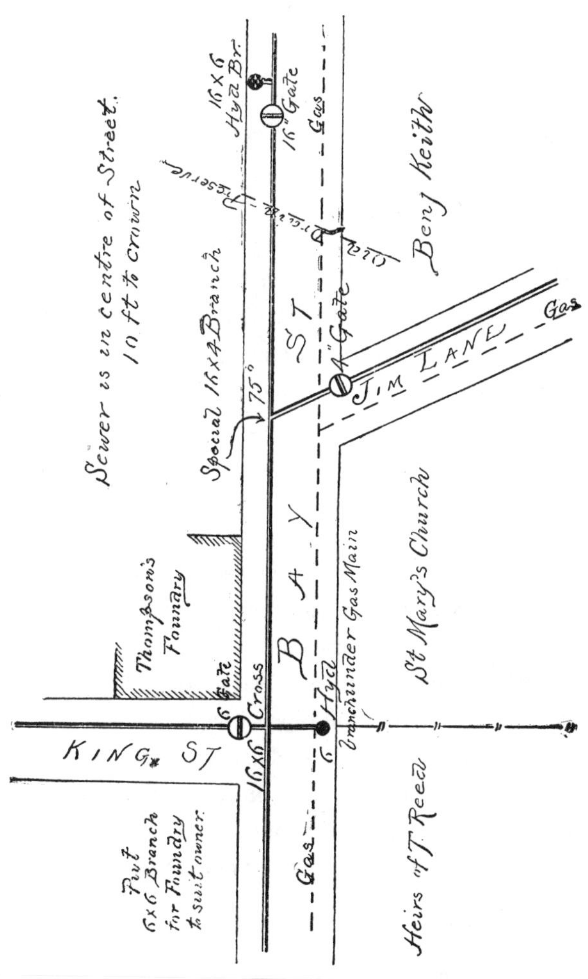
Sewer is in centre of Street.
10 ft to crown
Special 16x4 Branch
75°
16x6
Hyd Br.
16" Gate
Gas
Benj Keith
4" Gate
JIM LANE
Gas
BAY ST
St Mary's Church
Thompson's Foundry
6 Gate
Cross
16x6
Hyd
6
Branch under Gas Main
KING ST
Put 6x6 Branch for Foundry to suit owner.
Gas
Heirs of T Reed

FIGURE 11.

We must have a line of some kind—sash-cord or clothes-line are first-rate for the purpose—so that we can fasten one end of it to a point in the ground nearly over the branch, say 18 inches to the right of it. Now let a man take the ball or coil of line and stretch it in the direction in which the branch points, as nearly as can be judged. Suppose the line is to swing to the right, let one laborer drive his pick into the ground close to the line, on the right of it, and 12 or 15 feet from the fastened end. Keep the line stretched, swing it to the right again, and have another pick driven into the ground 25 or 30 feet from the fastened end.

When a hundred feet or more of the line have been stretched in this way, set a half a dozen men to picking a rut along the left side of the line. Make them follow the line; don't let them walk backwards, and see that they all pick on the same side of the line; it is safe to say that half of them won't if you let them alone. The sections may be measured by laying down a shovel-length four times, or, if the digging is sandy, 15 feet is not too much.

The width of the trench may vary from 28 to 36 inches, depending on the size of the pipe, though if the soil is known in advance to be sandy, and likely to cave, it may be cheaper to start the trench four feet wide on top, and slope it towards the bottom, rather than to use bracing.

The depth to which pipe may or must be laid is controlled by more than one consideration.

In northern latitudes protection from frost is first to be thought of, and the amount of covering required for this depends upon the nature of the ground, the size of the pipe, and the quantity of water flowing during the hours of minimum flow.

In loose, gravelly and stony ground the temperature will frequently fall below 32° Fah. for a depth of 5 or 6 feet, and hydrant-branches and service pipes have frozen under such conditions. In compact earth, free from large stones, the ground is not frozen more than three or four feet, and under good sod the distance is even less than that. These figures will hold good, I think, as far north as the isothermal line of Portland, Me.

Any section of a main pipe-line which is sure of a good circulation may be laid at any convenient depth without regard to temperature, and examples may be cited of main pipes which cross bridges without any protection from freezing except that afforded by the current of water constantly moving through them. Exact information upon this point is desirable.

There is a sort of unwritten law, in New England at least, that the axis of all pipes should be five feet below the surface.

If the amount of work on hand justifies the employment of not less than forty or fifty men, it will require the attention of one capable man whose duty shall be those of a foreman of the trenching gang.

The right man in this position will have no lack of work. He can keep the time for the whole gang, lay out the trench in advance, see that the damage from the excavated dirt is reduced to a minimum, keep private driveways open, look after the bracing of the trench if this be found necessary, see that the trench is dug to the line and grade given, keep the unoccupied side of the road as free as possible, and, finally pick out the fellows who are trying to shirk and get rid of them.

Time-keeping, if one wishes to know with exactness the cost of the whole or any portion of a season's work, is an

$2.25 Jack Cade 49	Total Time	August	Bay St	High St	Temp Sta
1	┌	DK.	┌		
2	□	"	□		
3	⊔	"	⊔		
4	┘	"	┘		
5	2.75	(Sunday)			
6	□	CK.	□		
7	□	" ½	┌	┘	
8	□	"		□	
9	□	"		□	
10	□	"		□	
11	□	"		□	
12	6.	Sunday			
13				□	
14					
15					
16	□	CK		□	
17	□	"		┘	┌
18	□	"		□	
19	3.	Sunday			
20	□	CK		□	
21	□	"		□	
22	└				└
23	┌				┌
24	□	CK		□	
25	□	CK½		┌	┘
26	5.	Sunday			
27	□ 2			□ 2	
28	□			□	
29	□			□	
30	□ 1			11	
31	□			□	
	5.3				

$1.75 Jim Daley 50	Total Time	August	Bay St	High St
1	□	BH	□	
2	□	BH	□	
3	□	BF	□	
4	□	BF	□	
5	4.			
6	□	C.S.P		
7	□	"		
8	□	"		
9	□	BH		□
10	□	BH		□
11	□	Tr.		□
12	6			
13	□	Tr.		□
14	□	"		□
15	□	"		□
16	□	BH		□
17	□	BH		□
18	□	BH		□
19	6			
20	□	Tr		□
21	□	"		□
23	□	"		□
23	□	CSP		□
24	□	"		□
25	□	"		□
26	6.			
27	□	M.M.P ½		┌
28	□			□
29	□			□
30	□			□
31	□			□
	5			

FIG. 12.

important detail, and a convenient and well-designed time-book is almost indispensable to good results. A sample page from the time-book in use by the writer will illustrate one method which has been well tried and is not found wanting (see Figure 12).

With the aid of this book we have been able to tell with satisfactory exactness at the end of a season's work where every dollar of the pay-rolls has been expended.

For example, the page here given tells us that Jack Cade is number 49 in the gang; that he is paid $2.25 per day; that on the first day of August he worked only during the fore-noon, in the derrick-gang on Bay Street; on the second he made a full day in the same position; on the third he did not begin until the middle of the forenoon, finishing out the day. As pay-day comes once a week the space belonging to Sunday is utilized to put down the footing of total time for each week. On Tuesday, the seventh, Cade during the forenoon worked on Bay Street and on High Street after dinner, and Ck. shows that he was employed in calking joints. In Tim Daley's record B. H. stands for "bell-hole digging," B. F. for "back filling," Tr. for trenching, and C. S. P. (construction service-pipe) shows that Daley was taken from the main pipe gang on those days and sent to dig service-pipe trenches.

In working through the streets of a town, especially in the portions occupied by well-kept estates, it is well to remember that a man with a newly-painted fence or a bit of smooth grass-plot is very unwilling to allow gravel or clay to be thrown against his fence or on to his lawn, even if the street be narrow and the workmen cramped for room. A few hemlock boards do not cost much and may save considerable growling, for if they are judiciously placed against the fence they will protect both it and the lawn. Under these conditions, however, there is some danger, if the dirt reach nearly to the top of the fence, of straining the structure and throwing it out of line. If this happens the fence must be straightened and the bill paid.

When tunneling is impracticable driveways and cross-streets may be kept in constant use by opening the trench half way across the space, leaving just driving room, and then digging on as usual. When the pipe-line is brought up to the undisturbed portion, the last two or three joints may be made without waiting for others, then enough of the trench immediately filled to furnish new driving room, and the undisturbed portion dug out by the derrick gang in quick time.

Bracing, if done to any considerable extent, is expensive work, but as it is not right and does not pay in Massachusetts to expose men to risk of injury, bracing the trench is sometimes not to be avoided.

If the tendency to cave is only slight and the trench is not more than five feet deep, sufficient support may be given by single planks running along just below the edge of the trench and held in place by short pieces of 4x4 joist, which are cut a little longer than the distance across the trench between the planks, and then driven in place with sledges.

In loose gravel or sand this sort of bracing amounts to little or nothing, for the stuff will run out from under the planks and finally tumble everything into the ditch.

Water-pipes are seldom laid to a depth which requires the thorough bracing and sheet-piling of deep sewer-work, but a simple sketch and a few words of explanation will make plain the vital points involved in the construction of ordinary sheet-piling. After excavating to a depth of four feet, a trench which must go four feet deeper, in quicksand, for example, it may be braced as indicated in Figure 13. Lay the 4x6 stringers B along the bottom of the trench and put a 10-foot

FIG. 13.

plank between each end and the bank. Cut cross-braces C long enough to drive in hard, and then fix the top stringers T in the same manner; the next is simply driving plank to make the sheet-piling complete.

It is not always easy to cut sticks of just the right length to be used for cross-braces C, and screw-jacks are economical in time and labor if much sheet-piling is to be done. We may use short jacks and a piece timber shorter than the width of the trench by the length of the jack, or, in narrow trenches, jacks of sufficient length to enable one to dispense with a timber brace may be preferred.

The one thing needful to make sheet-piling thoroughly effective is to keep the ends of the plank as much below the bottom of the trench as is possible, and to this end each plank should be driven frequently if only a little at a time. If the ends of the plank are chamfered and pointed, so as to help to throw them back against the bank and sideways against the plank last driven, better work can be done than with square-toed plank. If the amount of driving is considerable it will pay to protect the ends of the planks by a wrought-iron cap. Driving is to be done with wooden mauls, six inches or more in diameter and twelve inches long, bound with rings of wrought iron.

CHAPTER III.

TRENCHING AND PIPE-LAYING.

Caving—Tunneling—Bell-Holes—Stony Trenches — Feathers and Wedges—Blasting—Rocks and Water—Laying Cast-Iron Pipe—Derrick Gang—Handling the Derrick—Skids —Obstructions Left in Pipes—Laying Pipe in Quicksand—Cutting Pipe.

A TRENCH which is troublesome on account of caving grows worse the longer it is open ; if, therefore, the trenching gang is a good distance ahead of the pipe-layers, and water and quicksand are found within two or three feet of the surface, it is wise to send the diggers ahead on to dry ground, or make some other arrangement, so that the last two or three feet in depth of the wet trench will not be opened until pipe can be dropped into it. When caving occurs in wet, heavy ground some warning of the impending trouble is given by cracks in the surface, running nearly parallel to the side of the trench ; but in sandy gravel the drop comes without warning and men may be seriously injured. In any case the tendency to caving is increased by the weight of the excavated material piled up on one edge of the trench, and, if circumstances will permit, it is well to keep men on the bank to shovel back the material as fast as it is thrown out.

In soil that will allow it, tunneling will often save the public and individuals much inconvenience by carrying the trench under crosswalks, driveways, and railroad crossings, and the only tools needed are the tunneling-bars, mentioned in the list of tools, and long-handled shovels. A little practice and boldness in this detail will give very satisfactory results.

With cast-iron pipe, when the digging is good and the trench stands up well, it pays to put three, four, or half a dozen men at work digging bell-holes ; that is, enlarged places in the trench, spaced so as to come about the joints of the pipe, and large enough to give a man room to swing his hammer and get at all parts of the joint without unnecessary fatigue. There is little or no danger of getting the bell-holes too large, and plenty of room for the calker will do not a little toward insuring tight and strong work. The bottom of the trench should be dug out eight or ten inches for a length of four feet beyond the joint, and the sides worked out on the same scale to give ample shoulder room. These directions will have a queer sound when one is trying to make joints in quicksand, and at such a time fixed rules amount to but little. No end of grit, plenty of hard work, with some little planning, will make joints in places that seem all but hopeless for the first half-hour.

In these cases, bell-hole digging and joint-making must be done together, and some suggestions upon this detail will be given later.

Neither stony nor rocky trenches offer any serious difficulties, and even in ledge-work it is simply a question of time and money. If the bottom of the trench comes in rock which must be worked out by drilling and blasting, the ledge should be cut away to a depth which will allow sand six or eight inches in depth to be spread upon the rock, in which the pipe may be imbedded. If boulders are encountered which are too large to be taken out by the derrick, they should be well cleared from the confining earth by digging before applying powder or dynamite; this gives the explosive a fair chance, and digging is cheaper than drilling and blasting. Large pieces may some-

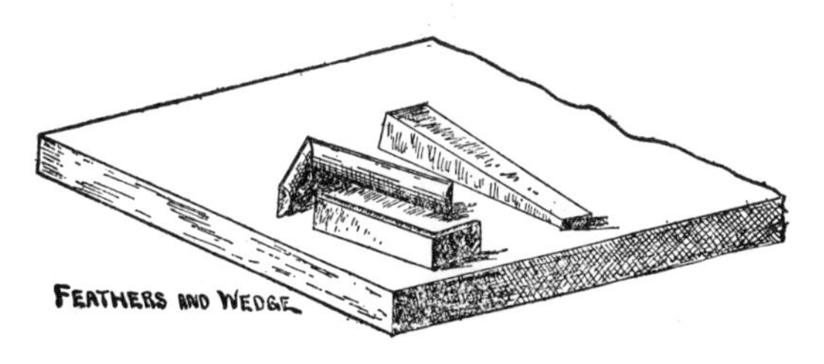

FIGURE 14.

times be worked off from a boulder or ledge which projects into the trench, without using explosives, by means of small hand-drills and "feathers and wedges." To do this, drill ¾-inch holes with a short steel drill and stone-mason's hand-hammer along the desired line of fracture, eight or ten inches deep and six inches apart; drop a pair of feathers made of ⅜-inch ½-round iron into each hole and drive the wedges between each pair. The "feathers and wedges" are shown in Figure 14.

In blasting, the nitro-glycerine preparation known to the trade as "forcite-powder" is comparatively safe and gives better results than common gunpowder, for it will shatter rocks more thoroughly and with less tamping. To fire a 1¼-inch hole three feet or more in depth, take a whole forcite-cartridge, cut off perhaps half an inch in length, and set a percussion-cap pinched on to the end of a piece of fuse into this short piece of the forcite by boring out a small hole with a knife. Lower this into the hole and cover it with the remainder of the cartridge broken into small pieces between the fingers, and fill up the hole with earth tamped down with a stick.

Such a charge as that will let daylight into any rock that a pipe-gang is likely to encounter, but the blast should be carefully loaded with logs, timbers, or railroad-ties chained together, and covered with brush to arrest small pieces which may do damage if allowed to fly. This forcite-powder may be used to loosen a troublesome boulder, by simply poking a hole into the bank alongside of it and tucking in a little of the explosive folded in an envelope and held in place by a slight packing of earth, or a cracked and seamy rock may be thoroughly split by dropping an envelope full of the powder into one of the cracks, and firing by cap and fuse in the usual manner.

Rocks which appear in the bottom of a wet trench are unwelcome enough, but it will not do to leave them in such shape that a pipe will be supported by them in the middle, with the weight of the back-filled earth bearing on the ends, lying in soft ground. If the expense of getting out the rock, seems too great, the depth of the trench should be reduced until a firm and even bearing can be secured.

On all trenches that do not stand up well or that must be made wide to get out rocks, the long three-legged derrick, illustrated on page 15, will be found exceedingly convenient, for its range is wide, and it can straddle fences in a right handy fashion.

PIPE-LAYING.

Cast-Iron Pipe.—When a hundred feet of trench has been bottomed out it is time to make up the derrick gang, and begin the work of putting the pipe into the ground. For six, eight, and ten inch pipe six men are enough, and they should be strong, active, and intelligent laborers. Men who are employed in this gang generally expect perhaps twenty-five cents per day more than the average digger, and good men in the place are worth it. It is not well to let the fellows who may be first chosen for this gang think that they are indispensable, and if one of them happens to be off a day, do not hesitate to take any good man out of the trench to fill the vacant place.

The first thing that a green lot of men must learn is to raise and carry the derrick, assuming that it be of the three-legged style referred to in a previous chapter. It is to be raised, first, just as a ladder should be, by footing the bottom and walking it into an upright position; then let one man grasp the pin of the middle leg with one hand and the leg with the other, a man at each of the other legs holding them firmly, and carry it straight away five or six feet; spread the other two legs the same distance, and the derrick stands alone, though perhaps not very firmly. A little study of the structure will now show that the legs may be spread as far apart as need be, provided always that lines joining the feet of the derrick form either an isosceles or an equilateral triangle, the line joining the two outside legs being the base. In placing over the trench, the middle leg should stand on the side which has the largest quantity of earth piled upon it. The man who is to carry the third leg, as the derrick is moved along from pipe to pipe, should grasp the pin firmly when the time for moving comes, throw his weight towards the trench, and be careful to keep midway between his comrades who are carrying the outside legs, and they in turn should walk as close to the edge of the trench as practicable, resist the push of the derrick firmly, and keep about ten feet apart.

A man at each leg, another to carry the rope, and two men in the trench, make an ordinary derrick gang; for handling 16-inch pipe more men will be needed in hoisting and placing. The smaller sizes of pipe can be brought from the side of the road to the trench by means of the carrying-sticks. These sticks thrust into a pipe give good lifting hold, and two stout fellows at each end, shoulder to shoulder, will carry 4-inch easily, and 8-inch without overwork. Skids of 4x4 spruce thrown across the trench may support the pipe while the derrick is put in place over it; a sling of rope is then to be passed around the pipe enough nearer to the bell than to the spigot end to cause the spigot end to fall easily into the trench when the pipe is lifted by the tackle from the skids. As the skids are removed to allow the pipe to be lowered into the trench, let one of the gang bunt the pipe with the end of the skid to clear the pipe from sticks, stones, and dirt. This is not enough, however, and it should be the duty of the men in the trench to look through the pipe as it comes down to them and make sure that no one has, either maliciously or carelessly, left therein an old hat, or a pair of boots or overalls. These remarks are not in jest, for just such combinations of what the doctors might call incompatibles have been made.

As the pipe is lowered, one of the trenchmen enters the spigot into the preceding bell, his comrade assisting as best he can, but before the pipe rests on the ground it is well to swing it like a ram against the pipe already laid to make sure that the joints ready for calking are all "home." As soon as the pipe rests on the bottom, the foreman should straddle the trench at a convenient point ahead of the derrick, *align* the pipe just laid, and look back over the line for joints which may be improved.

The trenchmen should carry bars with them to throw the pipe, and not try to use shovels for levers. Attention should be given to vertical alignment, as well as horizontal, and if grades are not given by an engineer, and no use is made of a carpenter's level on the pipes, the vertical alignment may be kept within bounds by keeping the joints of the same width at the bottom as at the top. If the bell end of a pipe when it rests on bottom is found to be too low, raise it with the derrick, throw rather more than enough loose dirt under it, and then drop the pipe down hard on this two or three times. As soon as the pipe is in position a few shovelfuls of earth should be thrown on to the centre of it to hold it, and if the trench is bad, the section between the joints may be half-filled at once, as this will support the bank and counteract any tendency to caving. With 4 and 6 inch pipe and a troublesome trench, two or three lengths may be put together on the bank, the joints made on dry land, and then with two derricks and careful slinging three lengths may be put into the trench at once without straining the joints. The few joints that must be made in the trench may, in quicksand, seem at first like hopeless cases, but persistence and no thought of ultimate failure have conquered the worst cases that have come in the experience of the writer. In such instances it is useless to attempt to get the sand down so as to make the joint right through without stopping to dig out again. Let the calker stand on the pipe while a good man with a shovel, perhaps a lot of sod, and some pieces of plank, clears away and holds back the stuff so that the joint may be yarned if not poured. If the sand rises as soon as the shoveling ceases, let the calker do all he can by quick work, and then rest while another attempt with planks, sod, pails, and shovels is made to make room for him. In general, whatever means are employed to make and maintain room for joint-making in quicksand, let the preparations be thorough; let the plank be driven as deep as possible and well braced, sods provided in large quantities; have pails or a good ditch-pump, and good strong men who are not afraid to "pitch in."

In order to locate gates or special castings in a particular spot, or to bring a joint into a more accessible location, it is frequently necessary to cut pipe.

For this use an 8 or 10 pound sledge and the long-handled cutting-off tool illustrated in Chapter I.; put a skid under each end of the pipe, placing one directly under the line of cutting and get a firm and even bearing on the ground for its whole length. A line for the cutter to follow may be had by winding the end of a tape-line about the pipe and marking along the edge with chalk, but a little practice will enable one to guide the cutter as the pipe is slowly rolled on the skids, so as to make a square cut. The blows of the sledge should be rather light for the first time around, and then when the cut is well marked so that it may be easily followed, the blows may be swung in with vigor.

The pipe should at some stage of the work be carefully inspected for cracks, which are oftenest found at the spigot end. If a crack in a spigot end is very slight and so short as to be more than covered by the bell, we may not think it worth while to cut the pipe, but a long crack obliges us to waste nearly twice its length of pipe, for the cut must be made at least six or eight inches above the visible end of the crack, and even then the jar of cutting may cause the crack to run still farther into the sound metal.

CHAPTER IV.

PIPE-LAYING AND JOINT-MAKING.

Laying Cement-Lined Pipe—"Mud" Bell and Spigot—Yarn—Lead—Jointers—Roll—Calking—Strength of Joints—Quantity of Lead.

CEMENT-LINED PIPE.

WROUGHT-IRON pipe after being lined with cement is not ready for immediate use. It should be allowed to dry for one or two weeks, the time varying with the weather, and the readiness with which the mortar sets, and a careful man will not subject the finished pipe in the trench to pressure for five weeks after laying, unless the pressure be very light. No derrick is needed in laying this pipe, for if circumstances do not allow the men on the bank to hand the lengths to their comrades in the trench as easily as they could lift a piece of stove-pipe, two pieces of rope will furnish means for easy lowering.

The cement bed and covering is "mud," in the language of a cement-pipe-laying gang, and is mixed sand and cement, three to two, in a mixing-box on the bank.

It may be conveyed in the trench in any convenient manner; in V-shaped troughs, ten feet long, with handles at each end, or in pails, or in wheelbarrows. Before placing a length of pipe, a bed of a dozen pailfuls of cement is spread along the bottom of the trench, thicker than the covering desired, and the pipe, with the rivets down, is pressed firmly into it; "mud" is then brought in sufficient quantities to allow the pipe to be plastered an inch in thickness, leaving the joints uncovered. The cement is spread with rubber mittens, and the men in the trench who handle the "mud" wear rubber leggings.

The joints are covered with pure cement, and are often made by the foreman of the pipe-layers, who can easily keep ahead of his men, for to a practiced hand the operation is simple and rapid.

The exposed pipe-ends are first covered with cement even with the finished pipe, and a sheet-iron sleeve is then slipped along so that its centre is directly over the joint. A pin of ¼-inch wire stuck into the trench will locate the butt joint of the two pipes, and make the placing of the sleeve an easy and certain matter, and the sleeve is then in turn covered with the pure cement.

This pure cement will crack, perhaps, and must be patched, and for this the regular "mud" will answer.

To protect the covering from too sudden drying, the pipe should be lightly covered as soon as it is laid, but the final covering should be delayed forty-eight hours.

The specials for cement-lined pipe can be made by any good sheet-iron worker.

Tee and Y branches are to be soldered at their junction and strengthened by knees of ¼-inch flat iron, one inch wide, riveted to the metal.

Plugs are simple cylinders filled solid with "mud," but they are to be braced in the trench with a heavy stone.

JOINT-MAKING.

There is not, to my knowledge, any standard form for a cast-iron pipe-bell or socket. This is unfortunate. The lack of agreement in this particular is, it is true, not nearly so unfortunate as the still greater lack of uniformity which prevails

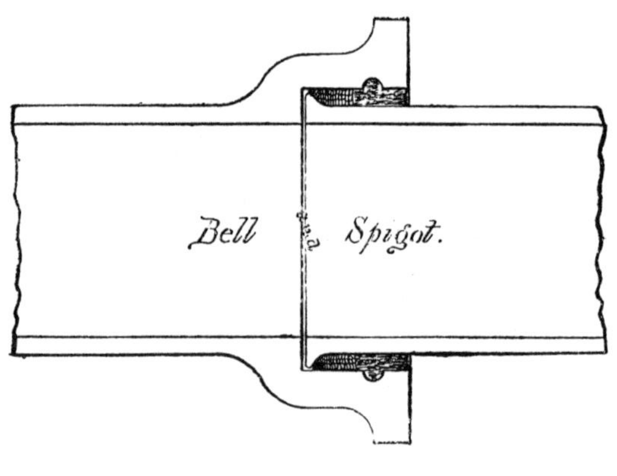

FIGURE 15.

in the thicknesses which are specified for cast-iron pipe, but a standard is desirable.

The general form of a bell and spigot pipe-joint is shown in Figure 15.

In practice, the two lines which in the sketch run through the word "and" should form one, and they will when the spigot end is pushed "home." A space is left in the sketch to make the parts more distinct. A water-tight joint capable of standing great pressure is secured by using a soft compressible substance in combination with molten lead. For the first substance one may use jute, hemp, old rope, old rigging, oakum, or almost anything of this nature, as the principal office of the "yarn," as it is oftenest called, is to prevent the molten lead from running into the pipe. It has been suggested that the yarn in the joints of a distributing system may, by its compressibility, serve to mitigate the shocks which come from the water-hammer, and again that the yarn will in time decay and may then furnish feeding-ground for noxious animal or vegetable life which may appear at one time or another in any water-supply.

At present these suggestions belong to that class of problems which are of special interest to the investigator. Of the first we may say that it has a reasonable appearance, and of the second, that if it be true the elastic cushion is lost when decay is complete.

The writer's experience has led him to adopt for yarn the article known to the cordage trade as 12-thread Russia gasket, tarred.

A larger size may be needed for 24-inch and 48-inch pipe, but the 12-thread has worked well on all sizes up to and including 16-inch.

For lead, use any soft pig, such as the "Omaha" or the "Aurora" brands.

In a gang of fifty, one man can find enough to do in yarning and pouring the joints. Let the yarn be cut into pieces long enough to go around the pipe and lap a little.

The yarner takes a bundle of these "ends" as large as he can conveniently carry from one bell-hole to the next, a couple of cold chisels, a yarning-iron, and a hammer, and, going to the first joint that is ready, he should, to begin with, see that the joint-room is even, or alike all around the pipe, and if it is not the chisels should be driven into the small places so as to crowd the pipes into line. This, of course, provided the pipes are intended to be in line, and one is not trying to get around a curve by "taking it out of the joints." The relative amounts of lead and yarn to be used per joint do not seem to be determined by any hard-and-fast rule. Referring to Figure 15 we can see that there is little except stiffness gained by putting in more than enough lead to reach back of the semi-circular groove, say one-quarter or one-half an inch, so that the depth and form of the bell must determine to a great degree the exact depth of lead in the joint.

Yarn is cheaper than lead, but the time consumed in yarning may, with lead at a very low figure, make it cheaper to put in only a shred of yarn and save time by filling up the joint with lead.

I think some contractors have figured in this way, for joints of their making which I have had occasion to dig up seem to have been made upon that principle.

Tarred stuff of some sort packs better and is easier to handle than dry rope or strings. The tarred Russia gasket, bought in 100-pound coils, is convenient to use for slings and lashings, and is just as good as ever for yarning after any other use. To guide the molten lead into the joint, we must have either a "roll" made of ground fire-clay upon a rope-yarn core, or a jointer. If a jointer is used, the yarner carries it with him in the trench, but a clay roll must be kept in shape and ready for use by the lead-boy. The patent jointers are made of canvas, rubber, and sheet-steel. They are very convenient, and can be obtained of dealers in water-works supplies. They are especially useful in wet places, for they do not easily blow out if a little steam is formed, and the clay roll will frequently give trouble in this particular. For making a good clay roll we require finely-ground fire-clay, a piece of board somewhat longer than the finished roll, a strand of rope, and a pail of water. Mix two double handfuls of the clay into dough, and after enough kneading to get out the lumps, roll the mass into a short thick club. With a stick or a chisel cut a slit lengthwise of the club and half-way through it, and lay therein a strand of rope a foot longer than the outside circumference of the pipe. Bring the two edges of the slit together, and then, by working, stroking, squeezing, wetting, and rolling, the roll may be drawn out to an inch in diameter, and eight or ten inches longer than the outside circumference of the pipe. This roll-making is the work of the lead-boy, who should keep the roll, when not in use, lying on the board covered with a wet cloth, and mend and wet it as the wear and tear demand. When he has packed the proper amount of yarn into the joint, the yarner should call out "roll" to the lead-boy, who will bring him the roll by the two rope-ends.

The roll is wrapped about the pipe close to the bell, bringing the two ends on top, and turning them out along the pipe, forming a convenient pouring-hole. The roll should be pressed firmly into place against the bell, and the molten lead poured in not too rapidly. The lead should be hot enough to run freely, and the furnace should be frequently moved, so that the hot lead need not be carried far enough to give it time to cool. After the joint appears to be full, and the roll has been removed, the yarner should examine the joint carefully all around, and especially on the bottom, to make sure that the joint *is* well filled; and if a cavity is found it should be filled by a second pouring if possible, or by a plug of cold lead. The calker follows, and should begin on the joint by

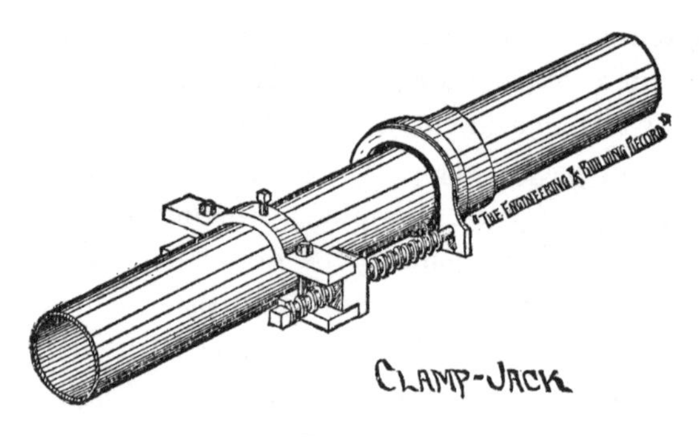

FIGURE 16.

using his chisel, cutting off the lump at the pouring-hole, and then driving the tool lightly between the lead and the surface of the pipe all around. Having, by this operation, lifted the lead away from the pipe, he begins with the smallest tool and drives back the lead, a little at a time, all round, and, following with the larger tools, sets the metal in firmly with strong, even blows.

Calking is hard work and needs a muscular man to follow it steadily, but it is not enough that he be—

> " Darbyshire born and Darbyshire bred,
> Strong in the arm and thick in the 'ed,"—

for he should know when a joint is right; but above all he must be trustworthy and faithful, and certain to call attention to any joint that he cannot get into proper shape without help. The quantity of power required to pull apart a well-made bell and spigot joint will surprise one who sees it measured for the first time.

In the experience which the writer has had in endeavoring to pull apart such joints the amount of force applied has not been measured with exactness, but a heavy clamp-jack having a pair of 1¾-inch screws with four threads to the inch, worked with a lever about thirty-six inches long, was insufficient to pull apart any but pipe from which the rim or bead on the spigot end had been cut off so as to leave a smooth end.

Some notion of the force applied to the joints by this clamp-jack, Figure 16, may be had by using the formula for power exerted by screw given in *Goodeve's Mechanics:*

$$P = \frac{w\,r}{a} \tan(\alpha + \theta),$$

in which

P = power applied at end of lever.
r = mean radius of screw-thread.
a = length of lever.
α = angle of thread.
θ = angle of repose.
$\tan\theta$ = coefficient of friction.
w = force exerted by screw.

Then.

$$w = \frac{P\,a}{r \tan(\alpha + \theta)}$$

In Figure 17, let B A represent the developed circumference of the cylinder on which thread is traced, and P A the

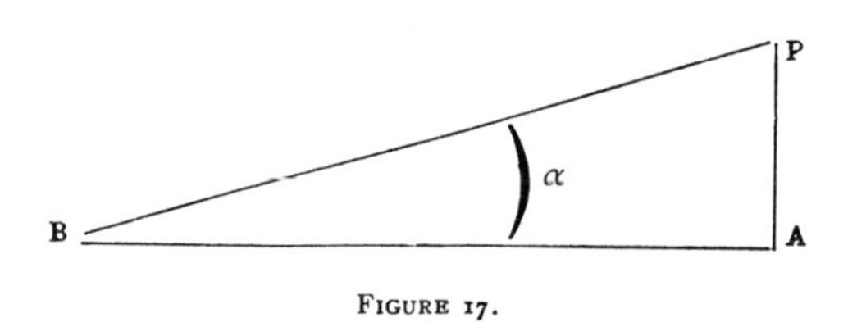

FIGURE 17.

pitch of the thread, and P B A $= \alpha =$ angle of thread. Then

$\tan\ \alpha = \dfrac{\text{P A}}{\text{B A}}$ and substituting the values for this case, calling 1½ inches the mean diameter of the screw thread,

$$r = \tfrac{75}{100} \text{ inch.} \quad \text{P A} = \tfrac{25}{100} \text{ inch.}$$

$$\tan\ \alpha = \frac{.225}{4.711} = .053 \text{ and } \alpha = 3^\circ\ 2'.$$

$$\tan\ \theta = .08 \text{ and } \theta = 4^\circ\ 35'.$$

$$\theta + \alpha = 7^\circ\ 37'. \quad \tan\ \alpha + \theta = .133724.$$

$$\text{P} = 100 \text{ pounds.} \quad a = 36 \text{ inches.}$$

$$w = \frac{\text{P}\ a}{r \tan(\alpha + \theta)} = \frac{3{,}600}{.75 \times .134} = 36{,}000 \text{ pounds.}$$

This formula makes no account of the power expended in overcoming friction at the pivot end of the square-threaded screw, and the result above given should be reduced 15 or 20 per cent.

The same clamp-jack has been found useful in pushing a hydrant off its branch for repairs.

As to the quantity of lead used in joint-making on cast-iron pipe the following notes are offered. Four streets having a total length of 3,112 feet of 6-inch pipe consumed 1,997 pounds of lead, or $\frac{64}{100}$ pound per running foot. Two streets, 1,796 feet of 8-inch required 1,514 pounds of lead, or $\frac{76}{100}$ pound per running foot.

During the past season the writer has directed the laying of 10,000 feet of 16-inch, 1,915 feet of 8-inch, 1,479 feet of 6-inch, 1,817 feet of 4-inch pipe. For purposes of this calculation it is fair to say that the quantity of lead varies directly as the diameter of the pipe, and that the above is equivalent to 11,927 feet of 16-inch pipe, and to make the joints on this 23,579 pounds of lead were used, or 1.97 pounds per running foot. This is larger, as of course it would be, than the amount given by a single experiment on a short piece, for ten pigs weighing 96.7 each (average weight) filled the joints on 550 feet of 16-inch, or 1.75 pounds per running foot.

The quantity of yarn used is not large, comparatively speaking, and on the three small sizes, 4, 6, and 8 inch, with the price at ten cents per pound, $\frac{6}{10}$ of a cent per foot is a safe figure for estimating purposes.

The quantity of pipe laid and the number of joints made in a day will, of course, vary greatly in different cases. If a man is trying to see how many pipes he can get into a trench, with the minimum amount of thought as to how they are put in and jointed, he can make a wonderful record, and the man who comes after him, and has to take care of the pipe-line under the shocks of service, will appreciate more keenly than any one else the value of such a record.

The following notes of actual work are offered, not in any sense as instances of model performance, but as simple illustrations: Time, July 6, 1887; gang 60 men, 16-inch pipe, 2 yarners, 2 calkers, 4 to 10 men digging bell-holes, 30 beil-holes per day, 400 feet of pipe laid and jointed in ten hours.

CHAPTER V.

HYDRANTS, GATES, AND SPECIALS.

STREET intersections are obviously suitable places for hydrants and gates.

A hydrant so placed serves more territory than one placed midway between cross streets, and at the intersection of important thoroughfares and large mains the four-way hydrants carrying four hose-nozzles are in every way suitable, if post-hydrants are chosen.

For the narrow crowded streets of a large city the flush hydrants are better than the post, but, as a rule, the small water-works which have sprung up all over the country during the last few years are fitted with hydrants of the post pattern.

If a post hydrant is not placed near a street corner, it is well to put it on a division line between two estates, for the chances that it will in the future be an obstruction are smaller in this position than they can well be in any other. The distance apart for hydrants may be 200 or 500 feet, according to circumstances, but the larger distance should not be exceeded without the best of reasons.

It has become a well-established custom to place gates on street lines, and the ease with which gates so placed can be found is a sufficient reason for not departing from the custom except in some special cases. In unpaved streets a gate-box located at a corner on a street line may be a source of trouble if the travel about the corner is considerable, for the wearing of the road will soon leave the box projecting above the surface to a dangerous extent. In cases where this condition of things is likely to obtain, the writer has thought it wise to move the gate ten feet away from the street line, and it is fair to ask if a uniform distance of ten feet would not have some advantages over a strict adherence to street lines.

The superintendent or the engineer or his assistant should follow the pipe-laying gang closely enough to locate every gate and special before it is covered by the back-filling gang. If one should perchance miss the location of something, he will be both surprised and amused to see how wild and yet how confident will be the guesses of a bystander who saw the gate covered the day before, and then tries to assist one in finding it.

In locating and making notes for future reference, a little judgment is required to enable one to choose permanent and easily-found landmarks.

Fences and stone-bounds come first, as a rule, and the post-hydrants furnish excellent measuring points. Lamp-posts are reasonably permanent, but trees and hitching-posts illustrate the "mutability of human affairs" of Dominie Sampson. A rough sketch, with no regard to scale, will be found more intelligible after sixty days than a written description.

As a rule, it does not pay to build gate-boxes so that a man can get into them to oil and pack the gates. In paved streets where digging is both expensive and inconvenient for the public, large brick manholes are of course demanded, but for town and country the cast-iron gate-boxes, well known to the trade, leave little to be desired.

The writer has heard of main-pipe specifications which called for a bed of concrete under each gate and hydrant. Under a hydrant in wet, uncertain ground the concrete may have some value, but under a gate there seems to be no call for it; indeed, it may be a source of trouble should the pipe settle a little and the gate be unable to follow. When a hydrant is placed in an ideal manner, it has a firm foundation in a large flat stone or good earth, good backing of stone or well-rammed earth and perfect drainage. If a sewer is not available, fair drainage may be secured by surrounding the base of the hydrant with broken or round stone, provided the ground has any absorbing power, and in clay, a small well may be sunk at some distance from the hydrant, enough below it and of sufficient diameter to contain three or four times as much water as the hydrant-barrel will hold. A small drain is then run from the hydrant to the well and the well is pumped out as often as need be.

Frost-jackets seem to be going out of fashion. Without doubt they have little value in sandy or gravelly soils. In clay the action of the frost may be expended on the jacket and so save the barrel some straining, but men of experience are not wanting who declare that the use of frost-jackets may be safely abandoned.

Generally speaking, the plugs for main pipe furnished by the foundries are unnecessarily heavy, unless made from special patterns.

In Figure 18 is shown the pattern adopted and used by the writer for the past five years.

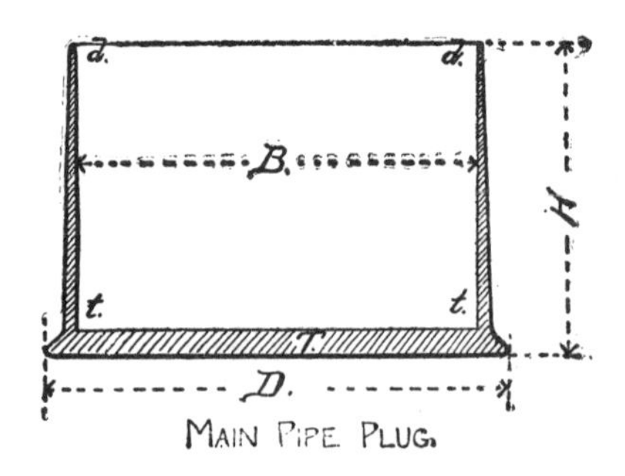

FIGURE 18.

The following table gives the dimensions for plugs to be used with four, six, eight, and ten inch pipe:

Size of Pipe.	D	B	H	T	*t*	*d*
4	5¼	4	6	⅜	¼	⅛
6	7¼	6	6	½	¼	⅛
8	9¼	8	6	½	¼	⅛
10	11¼	10	6	½	¼	⅛

The sleeve shown in Figure 19 differs from the ordinary pattern only in having an inside rim which furnishes a support against which the joints can be made. The diameter of this rim should be fixed with some care and with reference to the outside diameter of the pipe with which the sleeve is to be used; for unless the sleeve will slip over a pipe from which the spigot end has been cut, the chief advantage of this special casting will be lost.

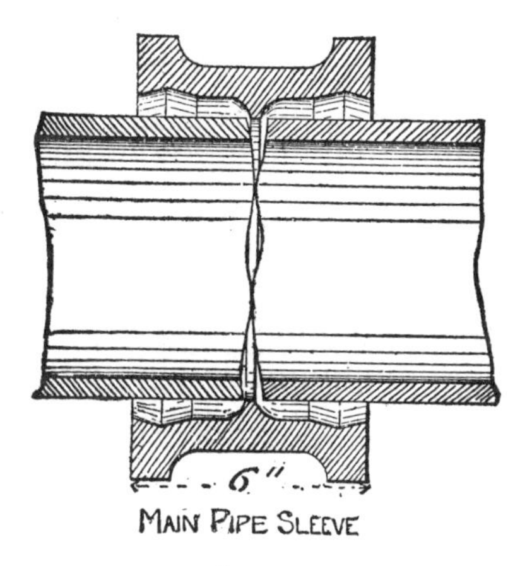

FIGURE 19.

Sleeves are all but indispensable in bringing two parts of a pipe-line to a junction between two rigid points, and they may be found useful in assisting one to use up pieces of pipe without bells. Some foundries make their special castings with bells all around, while others send out their single and double branches, with spigots on one end of the main run. The writer has found the "bells all round" pattern to be the most economical in the way of using up the pieces, but on every job of magnitude cases will arise in which the spigot-end special will save cutting pipe.

If practicable, main-line junctions should be made with specials a size or two larger than the pipe—that is, two 8-inch lines may cross each other at right angles, though a 10-inch double branch, and the New Bedford pipe plan by Mr. Coggeshall, given in a previous chapter, furnishes another case in point.

BACK-FILLING.

The best possible work in back-filling a trench is done with water, but oftener than not, perhaps, we must be content with ramming and tamping the dry earth. If time enough is put into it, and there is only one man shoveling to each man with a tamp, good work can be done without water, but such a method is expensive, and with contractors, as a rule, it is not in favor. The best results with dry earth are obtained when the dirt is spread evenly in layers, not more than six inches thick, and each layer is thoroughly tamped and trodden before another is added.

If he works as he should, the man in the trench will find the pounding and treading harder than shoveling, and to even things the shoveler and tamper may change places several times during the day. If water is used it should not be in such excess as to make "pudding" in the trench, and the amount of wetting must be proportioned to the absorbing power of the filling. The water does its work by carrying down the fine particles of earth as it soaks away, and more than enough to do this thoroughly is not needed.

If the trenching has been properly done, the top of the street—that is, the good gravel, or the macadam—has been put by itself on one side and should be raked over, and the stones and fine material separated ; the stones to be put in just under the surface which is to be finished with the fine material. The amount of crowning to be given the top of trench should depend upon the thoroughness with which back-filling has been done, the size of the pipe, and the character of the soil. If a trench has been well filled a rise of six inches is ample, and if this does not settle down even with the road after one or two hard rains it will have to be cut down if the road surveyor does not want to wait for wear and tear to level it. Some contractors prefer to fill without much tamping, crown the trench a

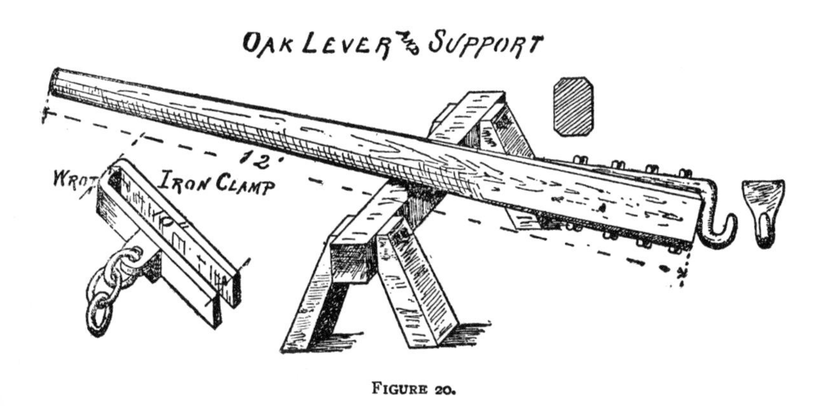

FIGURE 20.

foot, and then either repair the road after a month or two or deposit with the superintendent of streets a sum large enough to cover the cost of repairs. If sand has been taken from the trench it will ruin any road if allowed to come near the surface, by working up through a thin layer of good road material. If sheet piling has been used it may be removed after the trench is half filled by means of a clamp and lever shown in Figure 20. A **4x6** stick, a piece of chain, and a pile of blocks may be made to do the same work, but not so conveniently. The apparatus shown in Figure 20 is copied in part from a blue print presented at one of the meetings of the New England Water-Works Association by Mr. William B. Sherman, M. E., of Providence, R. I. The horse should be well braced with iron rods, and may be protected on top by a plate of light tank-iron.

FILLING NEW PIPES.

Pipes should be filled slowly and carefully, because under certain conditions great damage may be caused by too rapid filling. A long line should be filled one section at a time, and no gate before an empty section should be fully opened until positive evidence can be had that the section is filled. If the line to be filled carries hydrants, the air can be allowed to escape through them, but if these outlets cannot be had air-cocks on the summits are necessary.

A special form of air-cock can be had in the market, but for ordinary use any convenient form of corporation cock may serve the purpose by arranging a lever-handle and a blow-off pipe to be operated at will. In concluding the main-pipe division of his subject the writer presents in Figure 21 sketches of a tool wagon for use in main-pipe or sewer construction. The drawings are made from blue prints presented by Mr. R. C. P. Coggeshall, Superintendent of the New Bedford, Mass., Water-Works, at one of the meetings of the New England Water-Works Association.

TOOL-WAGON.

R. C. P. Coggeshall, Superintendent, New Bedford Water-Works.

This tool-wagon was planned by Mr. Ashley, foreman of this department, and was built by the regular employees during the winter months, at intervals whenever an hour or two could be spared. The cost as given

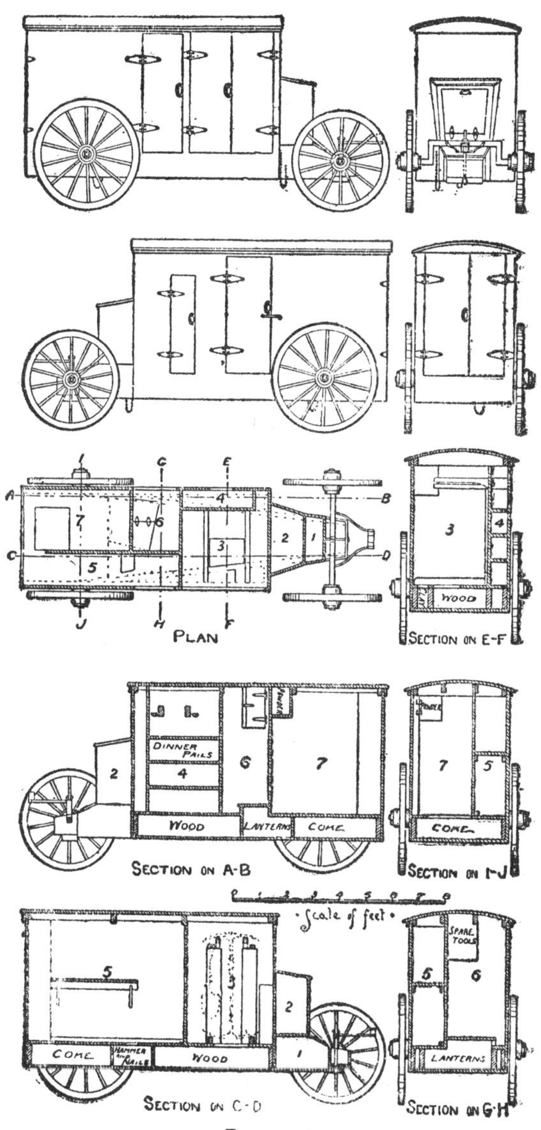
PLAN
SECTION ON E-F
WOOD
DINNER PAILS
WOOD
LANTERNS
COKE
SECTION ON A-B
COKE
SECTION ON I-J
Scale of feet
COKE
HAMMER AND NAILS
WOOD
SECTION ON C-D
SPARE TOOLS
LANTERNS
SECTION ON G-H

FIGURE 21.

below would in consequence probably exceed the amount at which this tool-wagon could be built by contract.

ESTIMATE OF COST.

Set of wheels and pole	$31 00
Axles, $10 ; bolts, $3	13 00
Door-pulls, 50c., 4 bolts $1.40	1 90
6 pair hinges, $1 ; 4 pair back-flaps, 40c	1 40
7 pair strap-hinges, $1.33 ; 1 dozen hooks, 60c	1 93
3 chain bolts, 90c.; 10 feet chain, $1	1 90
Screws, $4.16 ; nails, $2.15	6 31
303 feet 1-inch matched pine, planed	15 58
153 feet 1-inch matched spruce	3 52
130 feet 2-inch spruce, planed	2 33
Blacksmithing	30 49
Labor and painting	105 00
Amount	$214 36

CONTENTS.

1.

4 sets of lead and gasket irons,
4 drilling hammers,
1 stone hammer,
2 dozen cold chisels,
6 diamond points,
6 cutting-out irons,
12 joint wedges.

2.

4 lengths hose.

3.

40 picks and shovels,
3 stone sledges,
6 striking hammers,
Hydrant key,
Goose-neck,
Paving-pounder and hammer,
3 stone chains,
3 wheelbarrows of wood,
2 buckets of clay,
6-foot measuring-stick.

4.

20 dinner-pails.

5.

Tackle,
Nails and hammers.

6.

Small locker for spare tools,
Plug drill box,
9 lanterns and oil-can.

7.

Can, powder and fuse, 3 hoes, coil gasket, 6 pigs lead, furnace, 2 barrels coke, lead kettle and spoon, bell pole, saw, tamping bar, 12 buckets, 6 lantern sticks, 4 iron bars, 14 blowing-drills.

CHAPTER VI.

SERVICE-PIPES.

Definition — Materials — Lead vs. Wrought Iron — Tapping Mains for Services — Different Joints — Compression Union Cups.

BY common consent and general usage, the term service-pipe is applied to the tube which conveys water from the street-main to the premises on which it is to be used. In the majority of cases the service-pipe proper ends just inside the cellar wall, and the term house-pipes is a suitable one to apply to the tubes which convey the water from that point to the various fixtures in the building.

There seems to be substantial agreement among those best qualified to judge that lead is the most suitable material for service-pipes, but in spite of this the first cost of lead pipe and the popular prejudice which is often found against it has prevented its adoption in many recently constructed works. This is not the place for a thorough discussion of the subject, but those who care to follow it are referred to a paper by Mr. Walter H. Richards, C. E., Engineer and Superintendent of the New London, Conn., Water-Works, which was published in the transactions of the New England Water-Works Association for 1884, and to Professor Nichols' "Water-Supply from a Chemical Standpoint."

Lead pipe is to be preferred because it is the most durable, the most easily worked, and the smoothest pipe now in the market. Its substitutes are plain wrought iron, tarred or enameled wrought iron, galvanized iron, and wrought iron lined with cement.

One's choice really lies, then, between lead pipe and wrought-iron pipe with some protecting coating. Tin-lined lead pipe is not, to the writer's way of thinking, worthy of much consideration. The tin lining is thin and easily broken in working, and if the lead be exposed at any point the chance for some galvanic action, followed by the formation of lead carbonate or lead oxide, is too great to be taken. If any combination of chemical and physical reasons in some special case should render lead pipe unadvisable, a perfect though expensive substitute may be found in pure block-tin pipe.

The experience of every city and town which uses lead for service-pipe is, so far as I can learn, that a thin brownish insoluble coating soon forms on the interior walls of the pipe, and then all further action ceases. The cities of New York and Philadelphia; Boston, Worcester, New Bedford, Fall River, in Mass.; Denver, Col., Atlanta, Ga., Chicago, Ill., Wilmington, N. C., to go no further in this country, and Glasgow and Manchester abroad, use lead pipe, and this consideration would seem to dispose of the question as to its healthfulness, leaving only the question of cost to be considered, and upon this latter point Mr. Richards' paper referred to gives some interesting figures.

TAPPING.

Except for special reason, a main should not be tapped for service-pipes until it has been filled and, better still, if possible, not until it has been thoroughly flushed.

Cast-iron pipes must be entered by means of some sort of tapping machine. There are several machines for this work upon the market, and one will not make a mistake in buying any one of them, provided it is offered by trustworthy parties. It is well to bear in mind, in selecting a machine, that it is to be carried about, and perhaps knocked about; that it is to be used in all sorts of trenches, wet and dry, muddy, sandy, and rocky, and, therefore, that it should be light, strong, simple, and with as few wearing parts to collect sand and grit as possible. It will be well for any man who taps a pipe under pressure for the first time to choose, if he can, a section which can be easily shut off, for it will be nothing strange if he has to shut down and take off the machine to get the cock into the pipe. Printed directions for operating are furnished with each machine, and a week's work will make one independent of them.

That which is screwed, soldered, or driven into the main pipe is the corporation cock; at the sidewalk we have the curb or sidewalk cock, and just inside the cellar wall should be placed the house shut-off, or stop and waste cock.

In the early days of the Boston Water-Works sidewalk cocks were not used, and to shut off the premises wholly from the main the Water Department was obliged to dig down to the corporation cock. This condition of things was unsatisfactory, and, under the direction of Assistant Engineer Brackett, sidewalk cocks are being inserted.

As to the house shut-off just inside the cellar wall, there seems to be no good reason why the Water Department, or

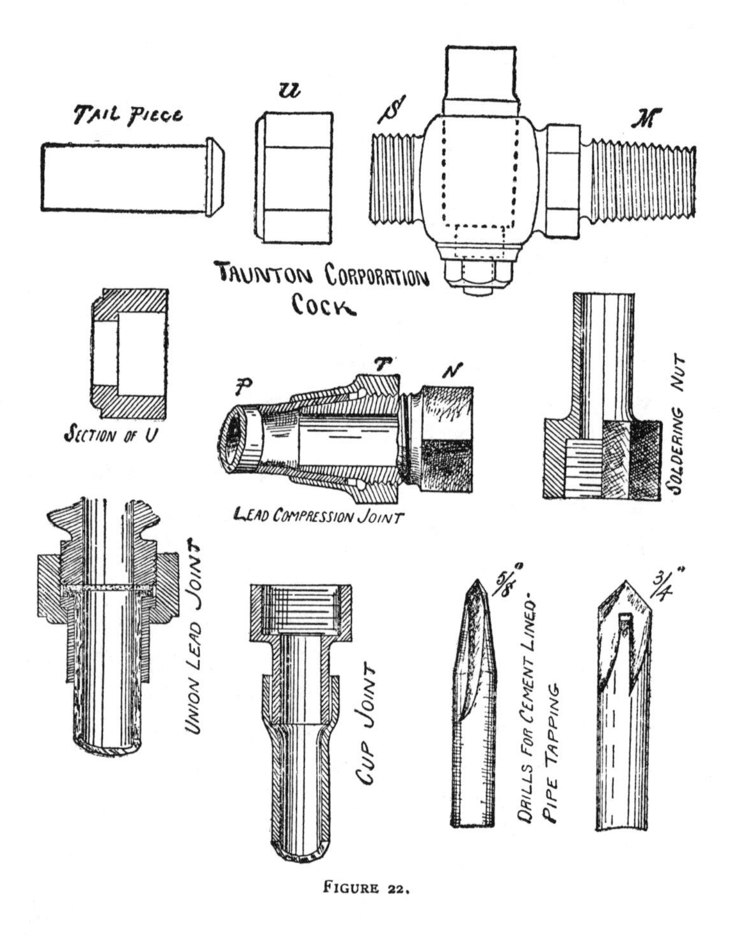

FIGURE 22.

the water company, should furnish that, except to secure uniformity and a first-class fixture. That there should be a good, sound, easy-working shut-off cock at that point there is no sort of doubt, but who should furnish it may be left as an open question. In Taunton it is furnished by the consumer.

Referring to Figure 22, in which is represented the particular pattern of corporation cock, with full ¾-inch way designed by the writer, for use upon the Taunton Water-Works, the end M is the end which is screwed into the main. The general form of this end is the same no matter what is used for service-pipe. Something is saved in the cost of manufacture by using the same thread at S and M. Eleven, twelve, fourteen, or sixteen threads to the inch are admissible, but fourteen has been found to give good results in the practice of the writer. At the end S and in the parts immediately following there is room for great variation in form and method. With the nut *u* (shown also in section) and the tail-piece forming a ground union-joint at the end S, this form of cock may be used (1) with lead service-pipe by making a wiped joint or a cup-joint between the lead pipe and the tail-piece; or (2) with any kind of wrought-iron service-pipe by joining on to the tail-piece a short piece of lead pipe, perhaps 18 inches, just as if the service were to be of lead pipe, and then, by attaching a soldering nut, as shown, and continuing the line with screw-joint pipe.

There is a form of corporation cock in the market in which the end S has a female connection so that wrought-iron pipe may be screwed directly to the cock without the intervention of lead pipe, but this form cannot be recommended for general use, because the flexibility of lead pipe is needed to insure safety against overstraining from settlement in the trench.

In addition to the joints made with lead pipe by wiping or cupping, there is one which may be called the compression-joint. Some regard this joint as to be preferred to any joint which depends upon solder, but the writer's experience does not lead him to take this view of it.

The compression-joint was in use a few years ago in Taunton, but was abandoned for a cup-joint. The corporation-cock then in use was shaped at the end S like the projecting part of N in the compression-joint shown in Figure 22, and tightness was secured by scraping the outside of the lead pipe to a reasonably smooth surface, so that the cone-shaped nut would draw the lead pipe firmly over the conical projection ; the lead pipe having been first spread by driving in a solid plug.

It is evident that this principle can be applied in a variety of ways, and that castings can be designed to fit any combination of materials. For example, the cup-joint in Figure 22 shows how a wrought-iron service-pipe may be joined to a lead connection from the corporation-cock. The lead pipe is attached by a wiped or cup joint to the soldering nut, which is tapped out to receive any size of wrought iron or brass pipe that one chooses.

Still another form of joint has been brought to my attention, by Mr. J. G. Briggs, Superintendent of Water-Works at Terre Haute, Ind., and shown also in Figure 22, as a union lead joint. Mr. Briggs says the idea is not a new one, but was used twenty years ago or more by an English company who did a large amount of work at Rio Janeiro, Brazil, and that in San Francisco the joint has been used for sixteen years with good results. The lead pipe is put through the brass thimble, and the end hammered or riveted over on a pin made for the purpose, and tightness secured by a washer. If this washer be of lead it will last, but it would seem as though a leather or a rubber washer would be too short-lived to be wholly satisfactory. As to the merits of this joint the writer has no practical knowledge, but the fact that Mr. Briggs favors it would, in the vernacular of the stock market, be counted as a "bull point" for it.

CHAPTER VII.

SERVICE-PIPES AND METERS.

Wiped-Joints and Cup-Joints—The Lawrence Air-Pump—Wire-Drawn Solder—Weight of Lead Service-Pipe—Tapping Wrought-Iron Mains—Service-Boxes—Meters.

THE regulation wiped-joint is one of the awful mysteries of the plumber's craft, and a description of its making would avail but little. It is the plumber's shibboleth, and if one of the trade can be found who will admit that any other joint is its equal he may be counted as one out of many. It is not to be denied that in many instances nothing can equal in appearance and fitness a well-wiped joint, and a thorough workman certainly knows how to make one; but a well-made cup-joint is equally strong—perhaps stronger—does not require a tenth part of the solder, and is made more quickly and with less practice.

A cup-joint is shown in Figure 22, and is made by expanding the end of the lead pipe with a properly shaped plug, scraping the inside of the cup with a jack-knife to give a surface of clean metal, dropping a soldering nut or tail-piece, properly tinned, into the cup, heating the whole joint by some appropriate method, and finally by filling the thin annular space between the cup and the tinned brass casting with melted solder. If these details are properly executed a perfect joint is the result. The writer has had several of these joints sawn in two and the bond is then seen to be perfect.

This joint was brought to the writer's attention by Mr. Dexter Brackett, Assistant Engineer of the Boston Water-Works, and a study of the method and its results will show that this is not a "tinker's joint," for it is used in Boston, Lawrence, New Bedford, and Taunton by the water departments of those cities, who have no sort of reason for using any methods or materials but the best.

The only portions of the process of cup-joint making which call for special mention are the method of heating the joint and the kind of solder to be used.

We should note in passing, however, that while the plug is being driven to form the cup, that this end of the lead pipe should be firmly held in a vise between two cast-iron half-round clamps that are cut out to correspond with the outside shape of the cup. When under these circumstances the plug is driven home, the lead forming the walls of the cup is compressed, and anything like a blister or defect has a chance of being closed.

For heating, Mr. Brackett uses, or did use, a sweating-iron, and so did the writer until Mr. Henry W. Rogers, formerly Superintendent of the Lawrence Water-Works, introduced a blow-pipe and air-pump apparatus, which is a great improvement in speed and convenience over a pair of hot irons.

The air-pump and the blow-pipe or lamp are shown in Figure 23. A jet of water, whose size may vary with the pressure under which it is to be used and the work to be done, from $\frac{1}{16}$-inch to ¼-inch, induces a current of air to enter the tee, and water and air together enter the separating chamber C made of 2-inch brass or iron pipe. The water flows off through the trap or bent pipe to waste and the air through the smaller pipe to the lamp or blow-pipe. When the apparatus is in operation the outlet for the air is so small that air accumulates in the separating chamber and forces the water down

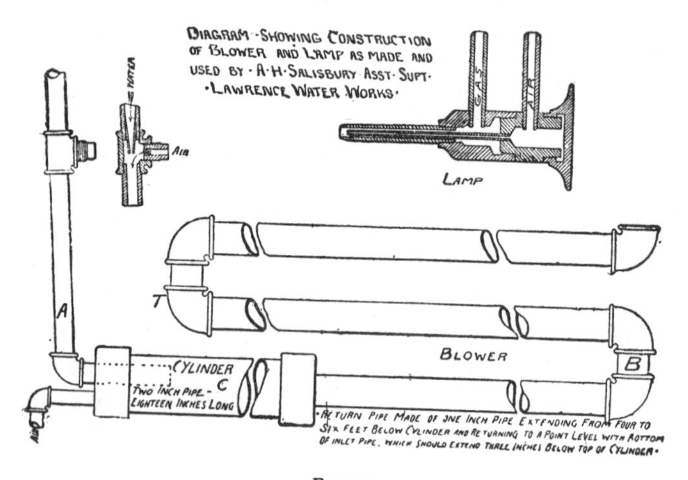

FIGURE 23.

below the the top of the trap a distance depending on the special conditions which exist in any given case; and the pressure under which the air accumulates is measured by the difference between the heights of the two water columns in C B and T B. The lamp is a Bunsen burner and the quantity of air from the pump, and of common gas from a convenient jet, may be so regulated as to produce a flame hot enough to make a bit of chalk glow like a calcium light. In fact there is an excess of heat for joint-making purposes, and a little experience will be required to prevent one from getting the metals so hot as to cause the solder to run through.

A very convenient form in which to use the solder is that given by drawing the common sticks into wire, about ⅛-inch in diameter. Wire solder has been for sale at a high price, and a large consumer would find it cheaper to build a small mill and draw the wire for himself than to pay twenty-five cents per pound.

There seems to be no standard weights for the various sizes of lead pipe, and an examination of a "Table showing weights of lead service-pipes used in various cities," which was compiled by Mr. William B. Sherman, of Providence, R. I., as an appendix to Mr. Richards' paper before referred to, will show more clearly than anything else the absence of uniformity.

For any but excessive pressures, exceeding 150 pounds per square inch, the following weights will be found sufficient:

Size..............Inches	½	⅝	¾	1	1¼	1½
Weight per foot...Pounds	3	3½	4	4½	5½	7

TAPPING WROUGHT-IRON MAINS.

There are more different methods of tapping cement-lined or coated wrought-iron pipe of any sort than of tapping cast-iron mains. Cast iron is seldom less than half an inch in thickness, but with wrought iron the actual thickness of metal is one quarter of an inch or less, and it is evident that such different conditions call for different treatment. Figure 24 shows in section the apparatus used for tapping wrought-iron kalamein pipe, used by Mr. Frank E. Hall, Superintendent of the Quincy, Mass., Water Co., and to whom I am indebted for a drawing of the machine. A packing of sheet-lead is put between the clamp and the pipe at the point to be drilled, and if tightness is not secured by screwing the nuts down hard, the lead can be calked up.

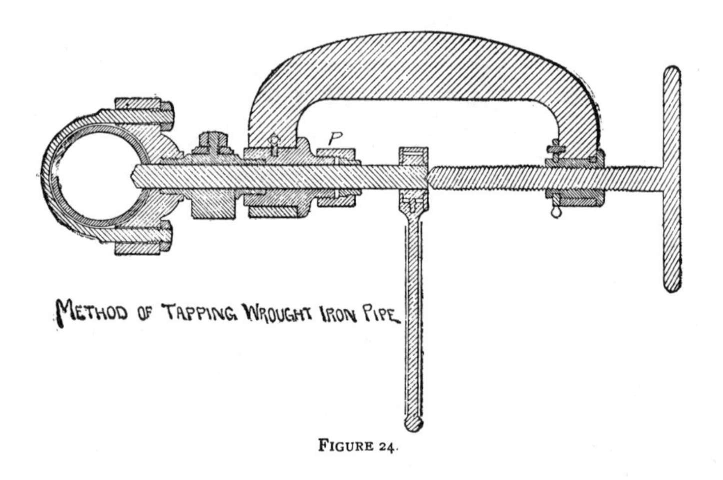

FIGURE 24.

With cement-lined pipe a similar clamp may be used, and such a clamp is a regular article of trade. A corporation cock may, however, be soldered or wiped directly on to the wrought-iron pipe without any clamp, and this is now the practice in many places. A small portion of the outside coating of cement is carefully broken away, the pipe is thoroughly cleaned and tinned, the cock is then attached to the main by soldering with an ordinary iron, or by wiping, and then, with an arrangement similar to that shown in Figure 24, a hole is drilled, passing the drill through the opened cock. After perforation the drill is withdrawn just far enough to allow the tapper to close the cock, and then the tapping apparatus is removed, the stuffing-box at P having kept the water back during the operation. Any convenient form of drill may be used, but Figure 22 shows the form used at Plymouth, Mass.

SERVICE-BOXES.

Considerable ingenuity has been expended in efforts to devise a cheap and satisfactory service-box. Wood was, naturally enough, one of the first materials to be chosen, and scored at first an apparent success when the stock was kyanized, but even if the preserving process proved to be in some cases successful, the frost made stumbling-blocks of the boxes by throwing them above the sidewalk level. Combinations of drain-pipe, light and heavy castings, and wrought-iron pipe with cast-iron bases might be described, but none of them, so far as the writer can judge, are any better, if as good as a simple cast-iron box in two principal parts sliding, telescope fashion, one inside of the other. The extension shut-off boxes, well known to the trade, give entire satisfaction, and at the price at which they are now offered it will hardly pay for any one to design a new pattern for any but special cases.

METERS.

Of making many meters there has been no end, and much experience with some of them is a weariness to the flesh. Of the six hundred or more that have been patented, six or less have come to any extensive use in this country, but in the value of that half dozen the writer has an abiding faith.

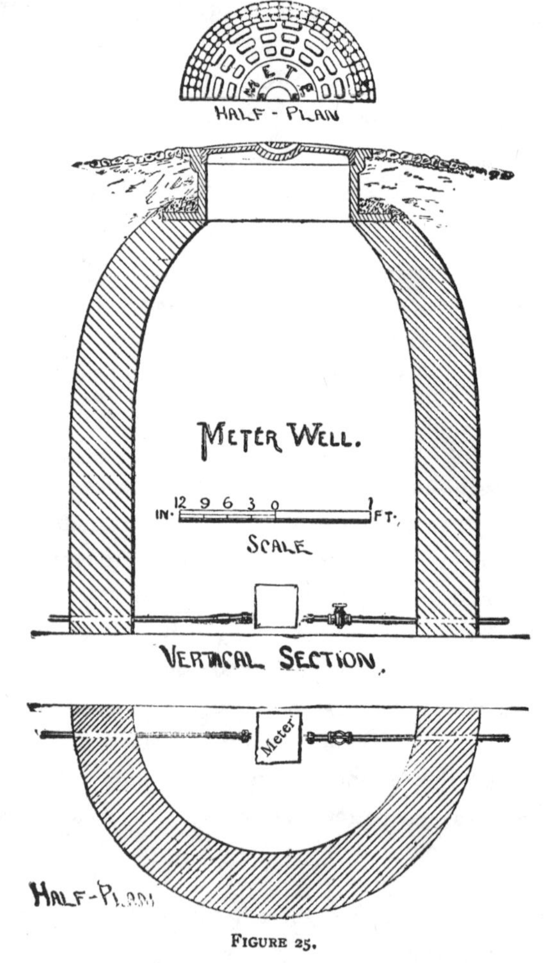

FIGURE 25.

The important points in setting a meter are perfect protection against freezing, a firm support, accessibility, and in some cases protection against meddlesome fingers. Too much care cannot be exercised in the first of these points, for a frozen meter is worth its weight in junk only, as a rule, and meters have been known to freeze in cellars in which the potatoes (so the owner said) never froze. If a house is to be metered, and the cellar is without a furnace, the safest place for the meter is just below the cellar bottom, and if the ground is too wet to allow this, and draining the cellar is out of the question, then a tight double box, with a 2-inch air-space, affords the next best solution.

Even if a house is not to be metered, it often is wise to enter the service-pipe from the street below the cellar bottom, as this affords protection to the pipe and secures cooler water in summer. In some cases the only place for a meter is in a driveway, a sidewalk, or a lawn, and in such cases a brick well with a cast-iron cover, the whole costing nearly $25, offers the best arrangement, as shown in Figure 25.

Meters should be well supported, either by a hanging shelf or a brick pier if one wishes to avoid all chance of springing the joints or the shell of a meter.

With lead pipe there is, of course, not the chance to hang the meter by the pipe that there is when iron or brass is used. It is quite important that a meter be so constructed as to have the inlet and outlet in the same line, and the distance from face to face of the inlet and outlet points exactly the same on all meters of the same size, for the best have to come out once in a while for repairs or cleaning, and then, with proper construction, a piece of pipe may take the place of the meter, with no inconvenience to the consumer.

CHAPTER VIII.

NOTES ON THE CONSTRUCTION OF ABOUT TWO MILES OF 16-INCH WATER-MAIN.

THESE notes are offered because the writer's experience has led him to believe that detailed statements of cost and of methods are not overabundant, and that a modest contribution to this department of engineering literature, even if it border on the commonplace, will not be unwelcome.

The city of Taunton, Mass., is supplied by direct pumping, and there is no store of water for any emergency. The pumping-machinery is in two portions, and under any ordinary conditions either portion is competent to maintain the supply, but we cannot, of course, be content with provision for nothing but ordinary conditions.

For that district of the city which is more distant from the pumping-station and higher than the City Square, the distributing portion of the system has been for some time inadequate, and, moreover, the small pipes have made it impossible for the city to receive from a powerful pumping plant belonging to a manufacturing establishment in that district the aid which might be rendered should the public pumping-machinery become disabled. The need for a new and larger main arising from the foregoing conditions was easily made evident to the proper authorities, and its construction was ordered.

The line was surveyed by the writer with one assistant in April, 1887, and the accompanying illustration shows the main in plan and profile, with its immediate connections.

The pipe began to arrive in May, and was carted on low two-horse trucks for 64 cents per gross ton, over good roads, for an average distance of about 1½ miles.

Referring to the plan, the work from A to B was without special features or difficulties. With the exception of a short stretch of quicksand and water at and near the first turn north of A, the digging was good and the trench required no bracing. The distance from A to B is 2,927 feet, and the cost of labor for this section was 32.3 cents per lineal foot. This includes all labor charged on the time-book, from the foreman to the water-boy in a gang of about sixty men.

From the point B to the end of the line at E, an 8-inch main was removed and a temporary supply maintained, so that no consumer on the line was without water for more than an hour or two at any one time. That the sections requiring temporary supply might be as small as possible, two gaps in the distributing system were closed ; the first one on Broadway north and south of Jefferson Street, between points F and G ; and the second between the dead ends on Pleadwell Street and on Fourth Avenue, which were brought to a junction, as shown on the plan. The first connection gave Jefferson, Madison, and Monroe Streets a continuous supply while Bay Street was cut off, and the second made possible a temporary surface connection, indicated by the dotted line, from Fourth

Avenue to Third and Fifth Avenues, which came in use when Whittenton Street was cut out.

It is to be understood, of course, that Washington Street continues in a northerly direction (see plan), and by cross lines completes a circuit for Whittenton, Bay, and adjoining streets.

The profile makes the proper positions for the blow-offs self-evident, and they are all six inches in size. The only portion of the pipe that cannot be drained by the blow-offs is found on Whittenton Street a few feet east of the line of the Old Colony Railroad, where a short trap exists, because of our unwillingness to disturb and wholly relay a first-class 10-inch drain which had been put in by the street department.

While this departure from the grade destroys the perfect drainage at which we had aimed, it will probably in actual practice be found to be of no real importance.

The position of the main 16-inch gates is shown on both plan and profile; they are of the ordinary upright bell-end Chapman pattern, not geared, with the exception of the one on Bay Street, near Maple Avenue, where the shallow trench obliged us to use a geared gate lying on its side.

At two or three points the stems of the upright gates came so near the street surface that the only box which could be used was Morgan's A A A extension valve-box, or one of like pattern.

The method followed in maintaining the temporary supply was adopted after careful consideration of three alternative methods; it is not new, for since this work was finished we learn that it is essentially the same as that followed by Mr. Coggeshall in a similar case in New Bedford.

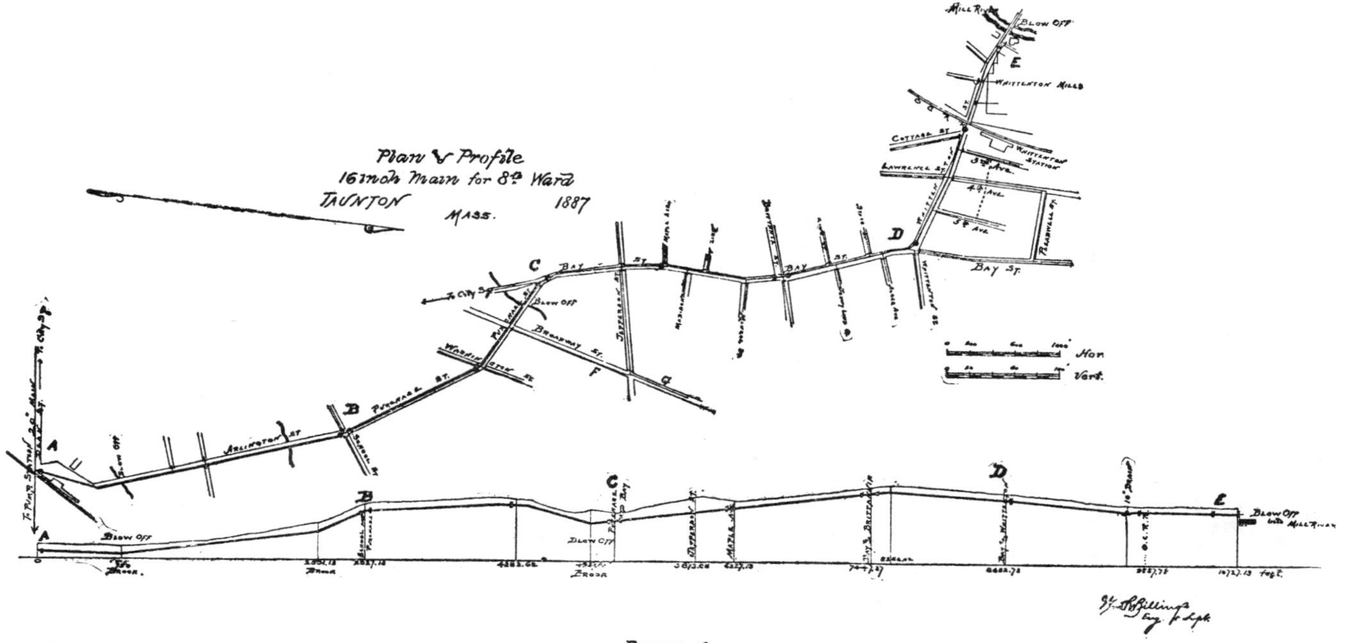
Plan & Profile
16 inch Main for 8th Ward
TAUNTON MASS. 1887
Arlington St
Purchase St.
Broadway St.
Jefferson St.
Bay St.
Cottage St.
Lawrence St.
Whittenton Mills
Whittenton Station
Pleasant St.
Blow Off
Hor.
Vert.
Blow Off into Mill River

FIGURE 26.

A temporary supply for consumers on a cut-out section may be furnished (1) by carrying water in tubs or buckets from the nearest available hydrant; (2) by laying a screw-joint pipe along the curb line, with stand-pipes at convenient intervals from which the consumers can draw at their pleasure; (3) by laying the screw-joint pipe as in the previous case, and then connecting each service-pipe by means of hose at a point near the corporation cock. The last method was adopted, and it is made clear by the accompanying sketch, in which the conditions represented are such that water under pressure comes as far as the large gate in the trench. The section which is temporarily shut off begins in front of the large gate and extends to the next gate on the old line, which is coming up, or to a point on the old line, which may be conveniently plugged if the next gate is too far away. The temporary pipe near the curbstone is common 1½-inch screw-joint pipe connected with the hydrant by 1½-inch hose and special brass couplings, and supplies 1-inch branches taken off at convenient points, carried down as shown, and connected with each service by ¾-inch 4-ply extra heavy rubber hose having special couplings, the nuts of which screw directly on to the end of the 1-inch cement-lined service-pipes, making a joint with a leather washer.

ITEMS OF COST.

Purchase Street.—In making preliminary estimates it is comparatively easy to get at the cost of materials, but the cost of labor and incidentals is oftentimes uncertain to an aggravating degree.

The following figures of cost of labor are believed to be as near the truth as it is practicable to get them without employing skilled clerical labor in keeping time.

Referring to the plan and profile it will be seen that Purchase Street for more than half its length is straight and practically level; it is forty feet wide, and, with the exception of a short section near Broadway, furnished sandy digging with some tendency to caving.

The crossing of the brook near Bay Street and the locating of the blow-off called for some comparatively deep digging—say ten or twelve feet in depth for 100 feet. The old 8-inch pipe was removed, and eighteen services were furnished with a temporary supply, and the total labor on this street cost $729.62. The distance is, say, 2,100 feet, so that the cost per lineal foot was 34.7 cents for the section between B and C on the plan.

Bay Street.—This though not the most expensive section was the most troublesome, for the difficulties were discouraging. The street is forty feet wide, has a horse-car track running through its entire length, with cars passing about once in fifteen minutes; from Maple Avenue to Britannia Street the line follows a sewer-trench so closely that the caving of the banks was almost constant. The digging was dry and sandy. The sidewalk on the west side was appropriated and all the excavated material was piled thereon; planks were thrown across the trench to enable the occupants of houses to pass in and out, and hemlock boards against the open fences kept the sand and gravel from the grass plots. The old 8-inch main was removed, the supply maintained for fifty-three services, the movement of the horse cars was not obstructed, and the total labor cost 41.8 cents per lineal foot—on section C D on plan.

Whittenton Street.—Here the digging was wet and dirty, but as the street is 65 feet wide there was ample room. Old pipe to remove, temporary supply to maintain for 30 services, and

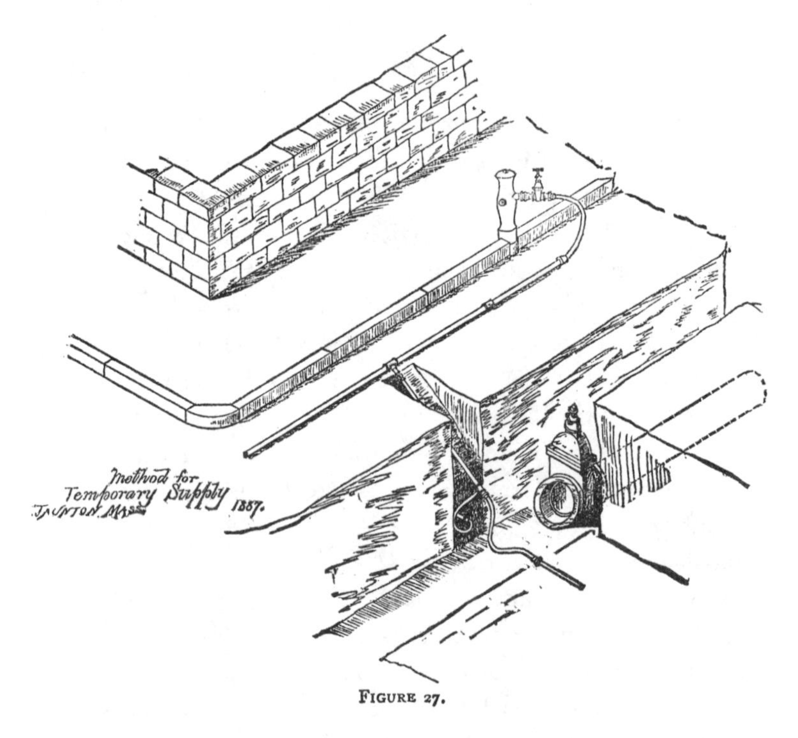

FIGURE 27.

four connections, new and old, made for the Whittenton Manufacturing Company at and near E on plan made the total cost of labor 47.4 cents per lineal foot. The mill connections were the principal causes of this increase in cost.

The foregoing figures are largely in excess of the cost of labor on ordinary pipe lines. For example, a detachment from the same gang of men who laid the pipe referred to above, laid about 2,000 feet of 8-inch pipe in new ground, good digging, at a cost of 17.3 cents per foot for all labor; two pieces of 4-inch, each about 530 feet long, for 13.1 cents per foot, and 600 feet of 6-inch for 15.38 cents per foot.

THE TEMPORARY SUPPLY.

The cost of work such as this will, of course, vary greatly with circumstances, for if new pipe and fittings must be purchased the cost will be much greater than it would be if old material and odd pieces can be worked up.

In this particular case we bought, expressly for this work, about half of what we used. The labor for the temporary supply-pipes footed up to $230.29, or about 3 cents per foot, while the new material purchased cost nearly 4 cents per foot in addition. A little less than 7 cents per foot for the 7,800 feet of pipe required for the temporary supply was the cost as nearly as can be ascertained. The pipe supplying Third and Fifth Avenues, from the hydrant on Fourth Avenue, was laid on the surface across the lots.

The total cost of the line and its connections may be stated as follows:

16-inch pipe, 135 pounds per foot (@ $34.50)	$22,698 09
6 and 4-inch pipe	563 52
Gates, hydrants, valves, globe special castings and sundries	7,728 56
Old style special castings	200 00
Labor	4,429 02
	$35,619 19

CHAPTER IX.

TABLES OF COST.

THROUGH the courtesy of Mr. Dexter Brackett, C. E., Superintendent Eastern Division of Boston Water-Works, I am able to present detailed dimensions of the cast-iron water-pipes used in that city, together with a table showing the cost of pipe-laying under Boston's methods and conditions :

Weights and Dimensions of Cast-Iron Water-Pipes, Boston Water-Works.

Diameter, in.	Class.	Dimensions in Inches.						Total length.		Total weight of pipes.	Weight per running foot laid.
		a	*b*	*c*	*d*	*t*	*l*	Feet.	Inch.	Lbs.	Lbs.
4	B	1.50	1.30	0.65	4.0	0.45	0.40	12	4	260	21.7
6	B	1.50	1.40	0.70	4.0	0.50	0 40	12	4	418	34.8
8	B	1.50	1.50	0.75	4.0	0.55	0.40	12	4	601	50.1
10	B	1.50	1.60	0.80	4.5	0.60	0.40	12	4½	815	67.9
12	A	1.50	1.60	0.80	4.5	0.58	0.40	12	4¼	935	77.9
12	B	1.50	1.70	0.85	4.5	0.65	0.40	12	4½	1,050	87.5
16	A	1.75	1.70	0.85	5.0	0.66	0.50	12	5	1,413	117.7
16	B	1.75	1.90	0.95	5.0	0.75	0.50	12	5	1,615	134.6
20	A	1.75	1.90	0.95	5.0	0.73	0.50	12	5	1,945	162.1
20	B	1.75	1.90	0.95	5.0	0.85	0.50	12	5	2,252	187.7
24	A	2.00	2.10	1.05	5.0	0.81	0.50	12	5	2,588	215.7
24	B	2.00	2.10	1.05	5.0	0.94	0.50	12	5	2,985	248.8
30	A	2.00	2.30	1.15	5.0	0.93	0.50	12	5	3,690	307.5
30	B	2.00	2.30	1.15	5.0	1.10	0.50	12	5	4,336	361.3
36	A	2.00	2.50	1.25	5.0	1.04	0.50	12	5	4,929	410.7
36	B	2.00	2.50	1.25	5.0	1.25	0.50	12	5	5,882	490.2
40	A	2.00	2.70	1.35	5.0	1 12	0.50	12	5	5,897	491.4
40	B	2.00	2.70	1.35	5.0	1.35	0.50	12	5	7,055	587.9
48		2.00	2.70	1.35	4.0	1.00	0.50	12	4	6,266	522.1
48		2.00	3.00	1.50	5.5	1.25	0.50	12	5½	7,917	659.7
60		2.25	3.40	1.70	6.0	1.375	0.50	12	6	10,959	913.2

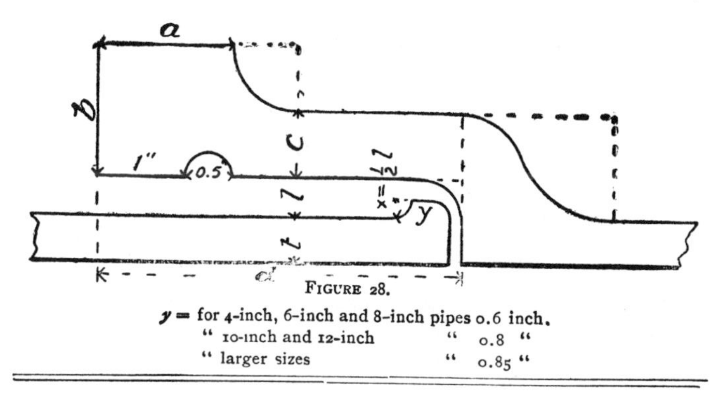

FIGURE 28.

y = for 4-inch, 6-inch and 8-inch pipes 0.6 inch.
" 10-inch and 12-inch " 0.8 "
" larger sizes " 0.85 "

"The pipe-joints are composed of hemp gasket and lead the lead being about 2½ inches in depth and thoroughly calked. The quantity of lead required for different sizes of pipe can be expressed by the formula $l = 2\,d$, in which l = pounds of lead per joint, and d = diameter of pipe in inches, and as the pipes are usually twelve feet in length, the quantity of lead required per lineal foot of pipe equals one-sixth of the diameter of the pipe in inches."

The average cost per lineal foot of water-pipe laid in Boston is shown in the table on page 92.

The centre of pipe is laid five feet below surface of ground. Labor at $2 per day. Pipe, 1½ cents per pound. Special castings, 3 cents; lead, 5 cents per pound. Cost of rock excavation, $3.50 to $5.50 per cubic yard, measured to neat lines.

By permission of Mr. Eliot C. Clarke, C. E., we are able to present the following useful tables of cost of excavation and brick-work. These tables, with others, were calculated especially for sewer-work, but apply, of course, to water-conduits as well, and the compilation of them was made for use during surveys made for the Massachusetts Drainage Commission in 1885:

Cost of Handling Water per 100 Linear Feet of Trench.

	5 feet Deep.	10 feet Deep.	15 feet Deep.	20 feet Deep.	25 feet Deep.
SLIGHTLY WET—Hand-pump...........	$6 00	$7 00	$9 50	$12 00	$18 00
QUITE WET—One steam-pump; one line 8-inch pipe at 20c. per foot; wells every 500 feet; moving engine, etc., every 500 feet; rent of pump and engine, $3 per day; one engineer, $2 50 per day; fuel..	71 50	73 50	76 50	103 45	127 45
VERY WET—Two steam pumps; 12-inch pipe at 36c. per foot; wells every 250 feet; two engines; three engineers; fuel.......	117 00	119 00	126 00	164 00	226 00

Average Cost per Lineal Foot of Water-Pipe Laid in Boston.

Diameter of Pipe. Inches.	Thickness. Inches.	Weight. Pounds.	Lead used. Pounds.	Cost of Pipe and Specials.	Lead Gasket and Blocking.	Teaming.	Labor, Trenching, and Laying.	Total Cost.
4	0.45	21.7	0.70	$0 38	$0 05	$0 02	$0 25	$0 70
6	0.50	35.	1.00	57	6	3	27	93
8	0.55	50.	1.35	83	8	5	30	1 26
10	0.60	68.	1.70	1 10	10	6	34	1 60
12	0.58-.65	78-88	2.00	1 27-1 42	13	7	37	1 84-1 99
16	0.66-.75	118-135	2.70	1 87-2 12	17	8	45	2 57-2 82
20	0.73-.85	162-188	3.35	2 55-2 94	21	9	55	3 40-3 79
24	0.81-.94	216-250	4.00	3 44-3 95	25	10	68	4 47-4 98
30	0 93	308	5.00	4 92	29	11	80	6 12
36	1.04	410	6.00	6 58	34	12	1 00	8 04
40	1.12	490	6.70	7 80	40	15	1 30	9 65
48	1.25	660	8.00	10 40	48	20	1 75	12 83

TABLE SHOWING THE COST OF CLOSE SHEETING PER 100 LINEAR FEET OF TRENCH.

Includes lumber, bracing, driving, and plank-pulling. Lumber used four times.

WIDTH OF TRENCH.	DEPTH OF TRENCH.									
	6 feet.	8 feet.	10 feet.	12 feet.	14 feet.	16 feet.	18 feet.	20 feet.	22 feet.	24 feet.
3 feet.........	$45 67	$55 65	$65 61	$74 10	$83 10	$98 10	$119 50	$146 13	$155 81	$165 50
4 "	45 80	55 80	65 82	74 75	83 40	98 60	120 25	146 53	156 23	165 95
5 "	45 93	56 10	66 20	74 95	83 75	99 10	121 05	147 19	156 79	166 50
6 "	46 06	56 35	66 60	75 35	84 12	99 60	121 85	147 85	157 40	166 95
7 "	46 19	56 60	67 00	75 95	84 50	99 90	122 40	148 51	158 01	167 50
8 "	46 32	56 85	67 40	76 50	85 00	100 25	122 75	149 17	158 77	168 40
9 "	46 45	57 12	67 80	77 10	85 52	100 60	123 40	149 83	159 80	169 75
10 "	46 58	57 40	68 20	77 55	86 05	101 15	124 10	150 49	160 25	170 75
11 "	46 75	57 75	68 75	77 95	86 60	101 65	124 80	151 25	160 80	171 85
12 "	47 10	58 20	69 30	78 20	87 10	102 25	125 50	152 00	162 60	173 20

TABLE SHOWING COST OF BRICK MASONRY IN EGG-SHAPED CONDUITS PER 100 LINEAR FEET.

Dimensions of Conduit.	3 feet × 2 feet.	3 feet × 2 feet.	3′ 6″ × 2′ 4″	3′ 6″ × 2′ 4′	4′ × 2′ 8″	4′ × 2′ 8″	4′ 6″ × 3′
Thickness.	4-inch.	8-inch.	4-inch.	8-inch.	4-inch.	8-inch.	8-inch.
Number of bricks at 25 in each cubic foot....	7,317	16,706	8,559	18,893	9,679	20,459	23,320
Barrels cement—1 cement; 1 sand..........	15.45	35.27	18.07	39.89	20.44	43.19	49.15
" " 1 " 1½ "	12.74	29.09	14.90	32.90	16.85	35.62	40.54
" " 1 " 2 "	10.84	24.75	12.68	27.99	14.35	30.31	34.49
Cubic yards of masonry....................	10.84	24.75	12 68	27.99	14.34	30.31	34.49
Cost of masonry at $10 00 per cubic yard.....	$108 40	$247 50	$126 80	$279 90	$143 40	$303 10	$344 90
" " 12 50 "	135 50	309 38	158 50	349 86	179 25	378 88	431 13
" " 15 00 "	162 62	371 25	190 20	419 85	215 10	454 65	517 35

Manholes, average about $3.50 for each foot in height. Iron covers and frames cost about $10 per set. Wrought-iron steps about 40 cents each.

TABLE SHOWING THE COST OF BRICK MASONRY IN CIRCULAR CONDUITS PER 100 LINEAR FEET.

Diameter of Conduit.	54-Inch.	60-Inch.	72-Inch.	72-Inch.	84-Inch.	84-Inch.	90-Inch	108-Inch.
Thickness.	8-Inch.	8-Inch.	8-Inch.	12-Inch.	8-Inch.	12-Inch.	12-Inch.	12 Inch.
Number of bricks................	26,812	29,673	37,402	54 979	40,142	62,829	66,758	81,091
Barrels of cement—1 : 1...........	56.82	62.65	78.96	116.1	84.75	132.6	140.1	171.1
" " 1 : 1½.........	46.86	51.67	65.12	95.73	69.89	109.4	116.2	141.1
" " 1 : 2...........	39.87	43.96	55.41	81.45	59.47	93.08	98.90	120.06
Cubic yards of masonry............	39.87	43.96	55.41	81.45	59.47	93.08	98.90	120.06
Cost at $10 00 per cubic yard......	$398 70	$439 60	$554 10	$814 50	$594 71	$930 80	$989 00	$1,200 60
" 12 50 " "	498 37	549 50	692 62	1,018 13	743 37	1,163 50	1,236 25	1,500 75
" 15 00 " "	598 05	659 40	831 15	1,221 75	892 05	1,396 20	1,483 50	1,800 90

In a letter to the writer Mr. Clarke says: "It should be understood that they (the foregoing tables) were made for a special purpose and are of limited applicability. Roughly approximate results were all we needed. Tables were based on then (1885) existing Boston prices for materials and labor, and average conditions affecting work." With this guiding statement the tables may be safely used in making preliminary estimates.